Combinatorial Development of Solid Catalytic Materials

Design of High-Throughput Experiments, Data Analysis, Data Mining

CATALYTIC SCIENCE SERIES

Series Editor: Graham J. Hutchings *(Cardiff University)*

Published

Vol. 1 Environmental Catalysis
edited by F. J. J. G. Janssen and R. A. van Santen

Vol. 2 Catalysis by Ceria and Related Materials
edited by A. Trovarelli

Vol. 3 Zeolites for Cleaner Technologies
edited by M. Guisnet and J.-P. Gilson

Vol. 4 Isotopes in Heterogeneous Catalysis
edited by Justin S. J. Hargreaves, S. D. Jackson and G. Webb

Vol. 5 Supported Metals in Catalysis
edited by J. A. Anderson and M. F. García

Vol. 6 Catalysis by Gold
edited by G. C. Bond, C. Louis and D. T. Thompson

Vol. 7 Combinatorial Development of Solid Catalytic Materials: Design of High-Throughput Experiments, Data Analysis, Data Mining
edited by M. Baerns and M. Holeňa

Series Editor: Graham J. Hutchings

Combinatorial Development of Solid Catalytic Materials

Design of High-Throughput Experiments, Data Analysis, Data Mining

Manfred Baerns
Fritz-Haber Institute of Max-Planck Society, Germany

Martin Holeňa
Academy of Sciences, Czech Republic

ICP

Imperial College Press

Published by

Imperial College Press
57 Shelton Street
Covent Garden
London WC2H 9HE

Distributed by

World Scientific Publishing Co. Pte. Ltd.
5 Toh Tuck Link, Singapore 596224
USA office: 27 Warren Street, Suite 401-402, Hackensack, NJ 07601
UK office: 57 Shelton Street, Covent Garden, London WC2H 9HE

British Library Cataloguing-in-Publication Data
A catalogue record for this book is available from the British Library.

Catalytic Science Series — Vol. 7
COMBINATORIAL DEVELOPMENT OF SOLID CATALYTIC MATERIALS
Design of High-Throughput Experiments, Data Analysis, Data Mining

ISBN-13 978-1-84816-343-0
ISBN-10 1-84816-343-6

Dedication

This monograph on the combinatorial development of catalytic solid materials is dedicated to all our friends and colleagues with whom we had many discussions and scientific exchanges on the subject particularly within the European network COMBICAT and at many international events. This dedication addresses in particular those scientists with whom we cooperated over extended periods of time with the expectation that this publication will be an inspiration to them and their collaborators to advance fundamentals and applications in combinatorial solid materials research: Laurent Baumes, Avelino Corma, David Farusseng, Cristina Flego, Michael Krusche, Claude Mirodatos, Carlo Prego, Ferdi Schüth, Selim Senkan, and José Serra.

Manfred Baerns and Martin Holeňa
June 2009

Preface

Combinatorial methods were first introduced in chemistry, for the discovery of pharmaceutical compounds and enzymes. Later on, the application was extended to homogeneous and heterogeneous solid catalysts. In this monograph, the combinatorial development of solid catalysts is dealt with. One of the authors, Manfred Baerns, first introduced an evolutionary technique using a genetic algorithm; in this way the composition of generations of catalytic materials was established. The work was initially done in cooperation with D. Wolf, O. Gerlach (formerly Buyevskaya) and then later with P. Claus, U. Rodemerck, D. Linke, S. Moehmel and N. Steinfeld. The second author, Martin Holeňa, joined the team in order to extend and deepen the application of mathematical and computer-science methods as prerequisites for the optimisation of the composition of the catalytic materials as well as for data analysis and data mining.

All these subjects are dealt with in this monograph and they are illustrated using some relevant case studies. In the final chapter, there is a collection of abstracts taken from relevant literature, which show the progress in the field along with some conclusions and an outlook on future developments. It is anticipated that the present text will enable more efficient execution of the development of heterogeneous catalysts.

The authors gratefully acknowledge the support that they have received from their former and present associates at the Leibniz-Institute for Catalysis at University of Rostock, Germany (formerly the Institute for Applied Chemistry Berlin-Adlershof, Berlin).

Martin Holeňa, acknowledges funding through the Grant Agency of the Czech Republic*.

Finally, special thanks are due the editorial staff of Imperial College Press, in particular to Sarah Haynes, who assisted us in many organisational matters and to the copyeditor who certainly improved the style of our original manuscript for this monograph.

Manfred Baerns and Martin Holeňa
June 2009

---

*Grant numbers: 201/08/0802, 201/08/1744, and ICC/08/E018

Contents

Chapter 1

Background of Combinatorial Catalyst Development

Catalysis has had considerable impact on the chemical industry for more than a century. In spite of its long-lasting practical application, scientific breakthroughs are still needed in terms of its fundamental foundations. This is true particularly for catalyst development, although significant contributions to the fundamentals of catalysis were recognized. Highly esteemed awards illustrated this only recently. In the field of heterogeneous catalysis various recognitions were awarded: the 1998 Wolf Prize to G. Ertl and G. Somorjai, the 2007 Nobel Prize to G. Ertl and the 2007 Priestley award to G. Somorjai. Also in homogeneous catalysis significant scientific progress was achieved as manifested through the 2001 Nobel Prize to W.S. Knowles, Ryoji Nayori, and K. Barry Sharp and the 2001 Wolf Prize to Ryoji Nayori.

For a full understanding of heterogeneously catalysed reactions the relations between synthesis and characterization of catalytic materials, and their catalytic performance with respect to activity and selectivity are required (Hinrichsen *et al.*, 1997; Ertl *et al.*, 1983, Ertl and Huber 1980; Grass 2008). Within the 21st century the main focus will be mostly on selectivity from a practical point of view for saving chemical resources and for avoiding harm to the environment from undesirable and poisonous products.

For an *a priori* design of catalysts only during the last decade, a few promising results have been obtained (Vang *et al.*, 2005, Neurock 2003; Jacobsen *et al.*, 2002); this is, in general, still impossible for multi-step reactions, for which selectivity plays the major role.

The industrial application of catalysis includes chemicals production and refinery operations as well as environmental protection (e.g. towards clean air and clean water) and energy-related (e.g. hydrogen production) processes; the latter two applications of catalysis are recently receiving increasing attention. In producing chemicals, usually at least one reaction step is driven by catalysis. About 80% to 90% of all chemical processes are based on catalysis and approximately 90% of this proportion involves the application of heterogeneous catalysts. Bulk chemicals are produced via either single- or multi-step chemical reactions in the refining and chemical industry including the production of primary chemicals for the synthesis of pharmaceuticals. Single-step processes include the Haber-Bosch process for the synthesis of ammonia, the oxidation of ammonia to NO as an intermediate towards nitric acid, and the oxidation of SO_2 to SO_3 for obtaining sulphuric acid. Multi-step processes, which consist of parallel and consecutive reaction steps, are among others reforming and cracking of crude oil-derived hydrocarbons, desulphurization, selective oxidation of alkanes and olefins to their oxygenates, hydrogenation of unsaturated hydrocarbons, and the synthesis of acrylonitrile and acrylic acid and of fine chemicals. Catalyst applications for environmental protection include those for automobile exhaust-gas purification, and for energy-related purposes such as hydrogen production via electrochemical catalysis for energy storage from solar plants. These selected examples illustrate the variety of heterogeneously catalysed chemical reactions.

The estimated annual value of all solid catalysts produced amounted in 2006 to about 13 billion euros per year worldwide and is expected to increase (not accounting for the present financial and economic crisis) to 16 billion euro in 2010 (Fischer 2009). The value of the catalyst market may be roughly subdivided to its different applications as shown in Table 1. The value created by these catalysts is in the order of about 100 to 1000 times as high. (Ertl *et al.*, 2008).

The commercial success of catalysis has been quite impressive and one might expect that catalysis is a mature science since it has now been industrially applied on a large scale for more than a century. However, this is not the case, as has been indicated above. Today, the *a priori* design of a catalyst for high product selectivity of complex (multi-step) reactions is mostly not yet possible. Fundamental knowledge still needs

Table 1.1. Approximate values of catalyst market (Fischer 2009).

Area	% of market value	
	2006	2010
Environment	33	(39)
Chemical and petrochemical	24	(19)
Refining	27	(18)
Polymerization	22	(24)

to be supplemented by empirical experience for discovering new catalysts for such purposes in a rather time-consuming procedure (see Chapter 2). However, one may anticipate that in about one to two decades, i.e. by about 2020 to 2030, it should be possible to predict the performance of a catalytic material in many situations on the basis of theory-driven approaches and accumulated fundamental knowledge.

At present, progress in catalyst design still very often relies on trial and error, even when partly applying fundamental knowledge. Therefore, statistical and machine-learning approaches are used to put the development process on a reproducible and rational basis. Such methods require a multitude of properly designed and collected experimental data for statistical analysis and deriving quantitative relationships between catalytic performance and the physical and chemical properties of the catalysts. Screening large amounts of materials, derived from combining different components, for their catalytic performance, requires suitable high-throughput technologies to obtain numerous, sufficiently reliable catalytic data over a limited time span by applying high-throughput technologies for preparation and testing.

Apart from experimental and dedicated theoretical methods, combinatorial computational chemistry approaches also exist (Koyama *et al.*, 2007), which are based on *ab initio* molecular orbital calculations. More details about these methods have been recently briefly summarized (Baerns and Holeňa 2008).

Combinatorial development of heterogeneous catalysts has proven to be a means of reducing the time for finding improved or new catalytic materials. The advantage of combinatorial catalyst development derives from appropriate, computer-aided design of experiments and their rapid execution using high-throughput technologies. Moreover, data evaluation and data mining play an important role in the combinatorial development of catalytic materials. These issues are the main focus of the present monograph.

It is, however, noteworthy that conventional methodology is still being successfully used in many cases (Trimm 1980; Stiles 1987; Le Page 1987; Richardson 1989; Becker and Pereira 1993; Wijngaarden *et al.*, 1998; Morbidelli *et al.*, 2001). A comprehensive and up-to-date overview of many aspects of preparation, characterization, and testing of solid catalysts, which is valid not only for conventional methodology but also for combinatorial catalyst development, has recently been published (Ertl *et al.*, 2008; Hutchings and Vedrine 2004).

Bibliography

Baerns, M. and Holena, M. (2008). Computer-aided design of solid catalysts, in Ertl, G., Knoezinger H., Schueth F. and Weitkamp J. (eds.) *Handbook of Heterogeneous Catalysis,* 2nd edition, VCH-Wiley, Weinheim, vol.1, 66–80.

Becker, E.R. and Pereira, C.J. (1993). *Computer-Aided Design of Catalysts*, Marcel Dekker Inc., New York, Basel, Hong Kong.

Hinrichsen, O., Rosowski, F., Hornung, A., Muhler, M. and Ertl, G. (1997). The kinetics of ammonia synthesis over Ru-based catalysts. Part I. The dissociative chemisorption and associative desorption of N_2, *J. Catal.* 165, 33–44.

Ertl, G., Prigge, D., Schloegl, R. and Weiss, M. (1983). Surface characterization of ammonia-synthesis catalysts, *J. Catal.* 79, 359–377.

Ertl, G. and Huber, M. (1980). Mechanism and kinetics of ammonia decomposition on iron. *J. Catal.* 61, 537–539.

Ertl, G., Knoezinger, H., Schueth, F. and Weitkamp, J. (eds.) (2008). *Handbook of Heterogeneous_Catalysis*, 2nd edition, VCH-Wiley, Weinheim, Vol. 1, Preface p. 5.

Ertl, G., Knoezinger, H., Schueth, F. and Weitkamp, J. (eds.) *Handbook of Heterogeneous Catalysis*, 2nd edition, VCH-Wiley, Weinheim, Vol. 1, Chapter 2.

Fischer, R.W., (January 2009). *Private communication* (Sued-Chemie AG Muenchen, Research & Development Catalysts, Bruckmuehl).

Grass, M.E., Zhang, Y.W., Butcher, D.R., Park, J.Y., Li, Y.M., Bluhm, H., Bratlie, K.M., Zhang, T.F. and Somorjai, G.A. (2008). A reactive oxide overlayer on rhodium nanoparticles during CO oxidation and its size dependence studied by in situ ambient-pressure X-ray photoelectron spectroscopy. *Angew. Chem. Int. Ed.* 47, 8893–8896.

Hagemeyer, A., Strasser, P. and Volpe, A.F. (eds.), (2004). *High-Throughput Screening in Heterogeneous Catalysis*, Wiley-VCH, Weinheim.

Jacobsen, C.J.H., Dahl, S., Boisen, A., Clausen, B.S., Topsoe, H., Logadottir, J.K. and Norskov, J.K. (2002). Optimal catalyst curves: Connecting density functional theory calculations with industrial reactor design and catalyst selection, *J. Catal.* 205, 382–387.

Koyama, M., Tsuboi, H., Endou, A., Takaba, H., Kubo, M., Del Carpio, C.A. and Miyamoto, A. (2007). Combinatorial computational chemistry approach for materials design: Applications in deNOx catalysis, Fischer-Tropsch synthesis, lanthanoid complex, and lithium ion secondary battery. *Comb. Chem. High Throughput Screening,* 10, 2, 99–110.

Le Page, J.F. (1987). *Applied Heterogeneous Catalysis*, Technip, Paris.

Morbidelli, M., Gavriilidis, A. And Varma, A. (eds.) (2001). *Catalyst Design*, Cambridge University Press, Cambridge.

Neurock, M. (2003). Perspectives on the first principle elucidation and the design of active sites. *J. Catal.* 216, 1–2, 73–88.

Richardson, J.T. (1989). *Principles of Catalyst Development*, Plenum, New York and London.

Schmidt, F. (2004). The importance of catalysis in the chemical and non-chemical industries, in Baerns, M. (ed.) *Basic Principles in Applied Catalysis*, Springer Berlin, 2–18

Selvan, P., Tsuboi, H., Koyama, M., Kubo, M. and Miyamoto, A. (2005). Tight-binding quantum chemical molecular dynamics method: A novel approach to the understanding and design of new materials and catalysts. *Catal. Today* 100, 1–2, 11–25.

Stiles, A.B. (1987). *Catalyst Supports and Supported Catalysts*, Butterworths, Stoneham MA.

Trimm, D.L. (1980). *Design of Industrial Catalysts*, Elsevier Scientific, Amsterdam, Oxford, New York.

Vang, R.T., Honkala, K., Dahl, S., Vestergaard, E.K., Schnadt, J., Laegsgaard, E., Clausen, B.S., Norskov, J.K. and Besenbacher, F. (2005). Controlling the catalytic bond-breaking selectivity of Ni surfaces by step blocking. *Nature Materials* 4, 160–162.

Wijngaarden, R.J., Kronberg, A. And Westerterp, K.R. (1998). *Industrial Catalysis, Optimizing Catalysts and Processes*, Wiley-VCH, Weinheim.

Chapter 2

Approaches in the Development of Heterogeneous Catalysts

2.1. Fundamental Aspects

As the science of catalysis has progressed the methodology of catalyst development becomes more rational and sophisticated, based on knowledge about catalysis and the fundamental requirements a catalyst should fulfil; also, experimental techniques, design of experiments and data analysis in the process of discovery of new catalysts have been significantly improved. From an applied viewpoint, catalysts that are more efficient than present ones are needed for many processes to reduce consumption of feedstocks and energy. This determines the direction of present catalyst development apart from the need for discovering new catalysts for novel reactions. In the development process emphasis is usually put on three aspects of catalyst performance in the order: selectivity > deactivation > activity.

For process engineering purposes, the type of reactor needed for a specific reaction usually determines the shape and texture of catalytic solid materials, which in turn, may influence inter- and intra-particle transport phenomena effecting catalyst performance. Most frequently, packed-fixed-bed, fluidized-bed, slurry-phase, and membrane reactors are used, which require different particle and pore sizes, shapes, specific surface areas, crushing and abrasion strengths (e.g. pellets, extrudates, spherical and granular particles, powders). Although these aspects play a vital role in the final preparation process for use of the catalysts in a pilot plant and later on in a commercial process plant, they are not discussed in this monograph, which focuses on the catalytic performance of a material. Selected references on the fundamental aspects of catalysts design have already been cited at the end of the preceding chapter.

In this chapter, general principles of catalyst design comprising composition (active components, supports), preparation, conditioning, and testing for screening of different materials are summarized. In this way, the reader is introduced to the challenges associated with catalyst development.

When designing a catalyst the molecular chemistry that is assumed to occur on the catalyst surface should be included in the process (see Table 2.1). Hypothetical reaction mechanisms have to be set up; moreover, the knowledge about the kinetics of the reaction steps might be helpful.

Table 2.1. Requirements in the design of heterogeneous catalysts.

Knowledge on molecular chemistry on the surface
Hypothetical reaction mechanism
Kinetics of reaction steps
Required surface properties for bond-breaking and bond-formation steps
Surface and bulk characteristics: elemental composition, acidity, basicity, structure, electronic properties (in particular, redox properties)
Types of active sites

In principle, all chemical transformations including bond breaking and bond-forming steps have to be considered; they require certain surface characteristics such as acidity, basicity, further, solid-state, electronic and in particular redox properties, and type of surface sites. With all this in mind and referring to empirical and theoretical pre-knowledge as well as intuition, patents, and literature, the catalyst developer starts carefully to think of possible components and support materials as well as methods of preparation for the catalysts.

On the basis of all the above variables and their interrelationships appropriate materials compositions have to be defined, prepared, and tested for their functional behaviour, i.e. catalytic performance. The choice of catalyst compositions (components and their proportions), in particular for complex (multi-step) reactions, which usually consist of larger numbers of components, becomes a challenging task considering that many elements and their compounds may fulfil such requirements.

That is to say, the optimal catalyst composition has to be searched for within a multi-parameter space, which should be covered to a large degree according to the strategies. This results in a large number of specimens (up to several thousand), which need to be prepared, tested for their catalytic performance, and finally screened for the best hits as precursors for further optimization. The design of the respective experiments is based on mathematical optimization methods (see Chapters 3 and 4). Data mining and data analysis serve as essential components of the development process, i.e. identifying the catalyst compounds as well as physical and physical-chemical properties, which significantly determine catalytic performance (see Chapters 5 to 8). In the authors' opinion, these techniques in combinatorial catalyst development may finally also contribute to increase basic understanding of the ongoing catalysis, which in turn may deliver new insights for further catalyst improvement.

For further details, compare Rodemerck and Baerns (2004) and Hagemeyer *et al.* (2004).

2.2. High-throughput Technologies for Preparation and Testing in Combinatorial Development of Catalytic Materials

In the extensive experimental work required in searching for an optimal catalyst, high-throughput technologies (HTT) are applied in the preparation and testing of the numerous catalytic materials. Simply put, HTT involves the fast preparation of large libraries of identifiable individual catalytic materials and their rapid physical and catalytic characterization.

HTTs have gained special attention in the search for and optimization of functional materials, particularly of solid catalysts since the end of the last century. Basically, HTTs are used for accelerating the development process for such materials comprising parallel synthesis, screening, and testing of a large number of materials applying heuristic methodologies. Such approaches include fundamental and empirical knowledge about the functionalities of the materials when progressing in the development process. As already indicated in the preceding section, it is obvious that in applying HTTs extensive experimental data are generated and accumulated. Appropriate data storage and data analysis are required for handling these data and for extraction of knowledge from them in terms of relationships between the chemical and physical properties on the materials and their functionalities, i.e. their catalytic performance. In industry, HTTs have led to the expectation of shortening development time for discovering new catalysts and hereby reducing the time-to-market of a new catalyst or even a new catalytic process. A conservative estimate is that the development time for a new catalytic material can be reduced to 10 to 20% comparing with conventional procedures Moreover, the ease and speed of preparing and testing new materials offers a unique opportunity to include elements which otherwise might have been left out of the development process. Thus, the probability of identifying new materials compositions is significantly increased.

The respective methodologies and dedicated experimental tools for solving the experimental tasks are described in the next sections.

2.2.1. *Selection of Potential Elements for Defining the Multi-parameter Compositional Space of Catalytic Materials*

The activity, selectivity and long-term stability of a catalyst depend on many variables, which define the multi-parameter space comprising chemical composition, mode of catalyst preparation and conditioning (calcinations, formation procedures) and reaction conditions (temperature, partial pressures or concentrations of reactants in the fluid phase being usually gas, but occasionally also liquid or gas/liquid systems). From this, it can be easily derived that a large number of variables describes the multi-parameter space in which the best catalyst exists. Accounting for catalyst composition alone, Senkan (2001), has calculated the number of possible catalyst compositions: Based on 50 chemical elements which might be considered for designing a catalyst, 1,225 binary 19,600 ternary and 230,300 quaternary combinations are possible. If different molar fractions of the components, different modes and parameters of preparation, and different reaction conditions are considered, a ‘combinatorial explosion’ occurs, in which case the number of experiments to be carried out would get out of control.

For starting any combinatorial development of catalytic materials for a specific reaction, a pool of elements/components has to be established from which — according to a predefined algorithm — a first generation of materials compositions has to be prepared. If different preparation procedures are applied, these could be also included in such an algorithm (see Chapter 3). For the sake of simplicity, at this point only compositional changes are considered. If, for example, an alkane is to be oxidatively dehydrogenated in the presence of gas-phase oxygen (ODH), assumptions have to be made about possible mechanisms of such a reaction. On this basis a pool of primarily selected components, in this case metal oxides, is set up. This is illustrated for ODH of ethane and propane to their respective olefins. Possible mechanisms for these two reactions are indicated in Figure 2.1.

For the three mechanisms different metal oxides were selected that differed slightly for the two reactions (see Table 2.2). For the first generation, randomized compositions with respect to type of component and its proportion in the material were prepared and tested for their

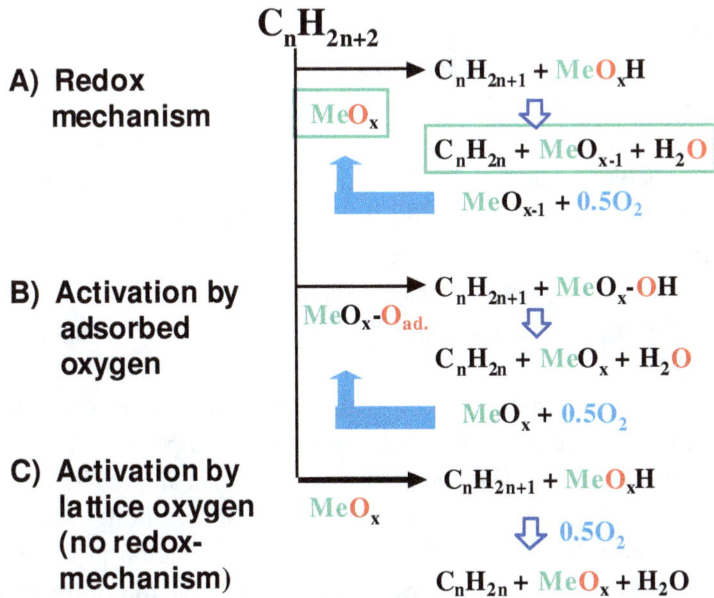

Figure 2.1. Selection of assumed primary reaction steps in the oxidative dehydrogenation of alkanes on metal oxides as catalyst components.

Table 2.2. Selection of a pool of elements from fundamental and empirical knowledge for preparing catalyst compositions for the oxidative dehydrogenation of ethane and propane (Rodemerck and Baerns, 2004; Grubert *et al.*, 2003; Buyevskaya *et al.*, 2000).

Assumed mechanism	Required property	Metal oxides	
		ethane	propane
Participation of removable lattice oxygen (Mars-van Krevelen mechanism	Redox properties (medium M-O binding energy)	Cr_2O_3, CuO, MnO_2, MoO_3,WO_3,Ga_2O_3, CoO, SnO_2	V_2O_5, MoO_3, MnO_2, Fe_2O_3, ZnO, Ga_2O_3, GeO_2, Nb_2O_5; WO_3, Co_3O_4, CdO, In_2O_3, NiO
Activation by adsorbed oxygen	Dissociative adsorption of oxygen	CaO, La_2O_3	La_2O_3
Activation by lattice oxygen	Non-removable lattice oxygen: high Me-O binding energy	ZrO_2	Acidity: B_2O_3 Basicity: MgO

functional properties, i.e. catalytic, performance. For further improvements in the performance of the materials tested, the compositions of the materials were changed according to a genetic algorithm; other optimization methods could have been also chosen; (for more details on this subject see Chapter 3).

Without going into further details, the results on catalytic performance — expressed as olefin yield at complete oxygen conversion — are shown in Figure 2.2. It is obvious that the yield for the best materials of each generation increases first with increasing number of generations until a maximal value is reached; (the average yield based on all catalytic materials of a generation with the number of generations behaves similarly).

Improvements might be possible via applying a conventional optimization of catalyst composition in the compositional range of maximal performance. The catalysts may be further improved by taking all experimental performance data into account by fitting them to an artificial neural network which describes the relationships between catalytic performance and composition over the whole multi-dimensional space (for details see Chapter 6). Some fundamental insights in catalysis may be eventually derived from such results.

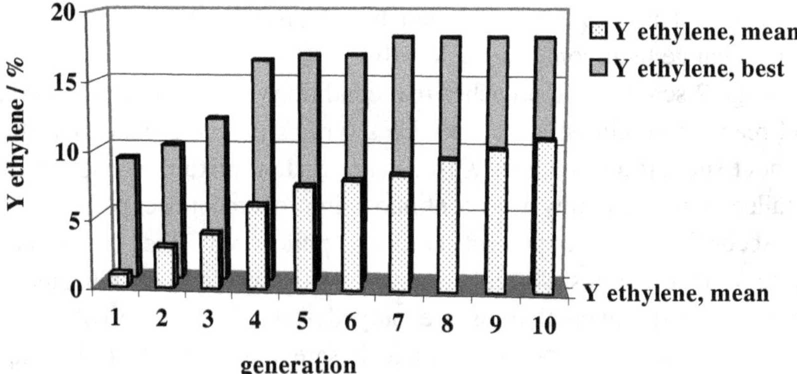

Figure 2.2. Example of catalyst optimisation applying a genetic algorithm for the oxidative dehydrogenation of ethane to ethylene: maximum and average olefin yield as a function of the number of generation in the iterative development process (after Grubert *et al.*, 2003).

2.2.2. *Experimental Tools for Preparing and Testing Large Numbers of Catalytic-material Specimens*

For covering the whole compositional parameter space, a combinatorial design of catalyst libraries — i.e. a full grid search — is required. The combinatorial design takes into account all elements at different randomized levels of concentrations. All catalytic materials from the libraries have to be tested for their catalytic performance, which can be quantified by different objective functions such as selectivity, activity, and stability. As already indicated in the preceding section, the required number of materials is usually very high depending on the number of elements involved and their concentration levels, which have to be defined by the experimenter. To a certain degree, the number of parameters can be restricted using available knowledge and intuition. The type of material desired determines the method used for preparing and testing. There are basically two forms of materials for catalytic testing, which fulfil different purposes, i.e. stage 1 and stage 2 screening (see Schunk *et al.*, 2004). Stage 1 screening features maximum sample throughput but delivers only reduced information, since only target compounds are analytically identified. This method is primarily used for discovering new materials, the compositions of which have to be further optimized. This procedure is often based on solid films on supports, which change their composition locally.

Stage 2 screening approaches real conditions for particulate materials and reaction conditions; solid particles (grains or micro-pellets) with or without support are frequently used. The reactant mixture is subjected to detailed analysis. In this way, continuous improvement is expected.

According to the present authors' philosophy, the evolutionary strategy incorporates stages 1 and 2 as a whole (see Chapter 2). Therefore, experimental tools are only described for the evolutionary strategy, which corresponds to stage 2 screening (for stage 1 sample preparation see, e.g. Cong *et al.* (1999)).

2.2.2.1. *Preparation of catalytic materials*

For high-throughput preparation of solid catalytic materials, automated robots equipped with synthesis control protocols are used. These robots carry out the various synthesis steps at high speed and in parallel or consecutively. The steps require high accuracy and reproducibility, without which no reliable data on optimal catalyst composition and best performance can be obtained. Moreover, the steps have to be performed in such a way, which is transferable to pilot- and technical-scale plant operation. For illustration, an automated synthesis robot is presented in Figure 2.3.

The algorithm, according to which the compositions of the individual specimens of catalytic materials are to be prepared, determines the synthesis protocol. If an evolutionary approach is applied the composition for the first generation is randomized and subsequent generations are synthesized according to a genetic algorithm including crossover and mutation (see also Chapters 3 to 4).

Another algorithm for saving effort in preparing a large number of catalytic materials is the split-and-pool approach — already known in combinatorial synthetic chemistry — which has been recently adapted to heterogeneous catalysis. This is shown schematically in Figure 2.4 (for details see Newsam *et al.*, 2000, 2001, Schunck *et al.*, 2004, and Rodemerck and Baerns, 2004).

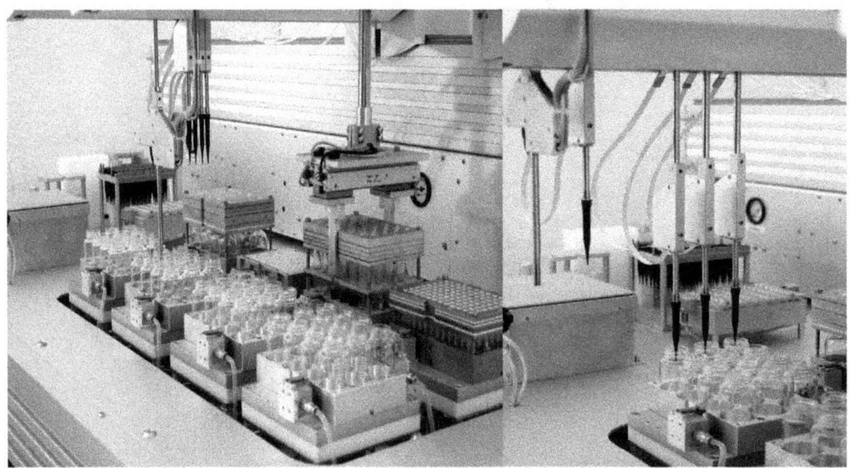

Zinsser Analytic

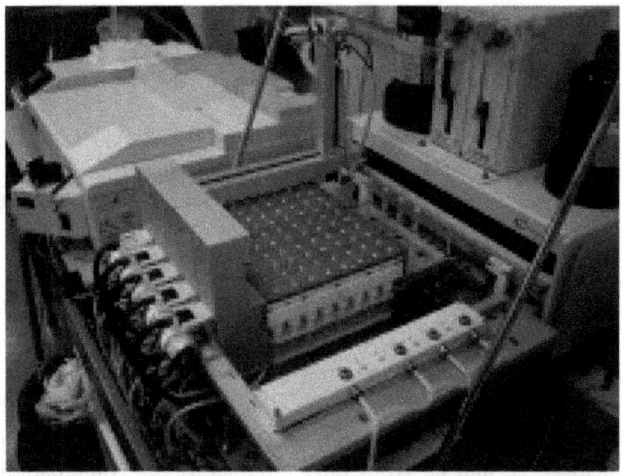

Chemspeed Technology

Figure 2.3. Automated high-throughput equipment for preparation of solid catalytic materials including dosing and mixing aqueous solutions of catalytic compounds, dosing of solid supports, precipitation of hydroxides from solution, filtering, washing and drying. (calcinations have to be done in a separate set-up). (Photographs courtesy of Chemspeed Technology, Basel, Switzerland and Zinsser Analytic GmbH, Frankfurt am Main, Germany).

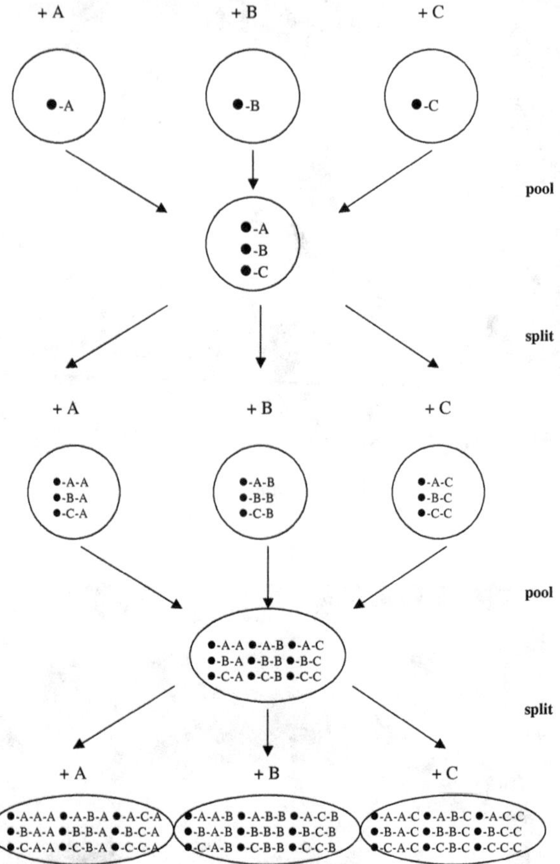

Figure 2.4. Principle of the split-and-pool method for preparation of large libraries of mixed-metal oxide catalyst: beads (black dots) are covered in liquid phase by the components A, B, and C, then they are separated from the liquid and united before being exposed again to A, B, and C. After only three steps, 27 combinations of components have been prepared.

2.2.2.2. *Testing and screening of catalytic materials*

For stage 2 testing of catalytic materials, fixed-bed reactors are used in parallel. The requirements for testing catalytic materials correspond to those for single-reactor operation: internal and external heat- and mass-transfer limitations must be excluded during testing to ascertain catalytic results, which are unbiased by transport phenomena. Furthermore, isothermal operation is needed, i.e. radial and axial temperature gradients have to be avoided. Appropriate criteria regarding these phenomena can be found in textbooks dealing with catalytic reaction engineering.

Comparable activity measures for each catalyst in the different parallel reactor channels can be derived from the degree of conversion of a key-feed component at the same contact time (catalyst volume divided by flow-rate of feed [volume/time]). On exposing all the open cross sections of the reactor channels to the total gas volume, it is evenly distributed to all channels as long as the pressure drops in the individual reactor channels are the same. If his is not the case, measures are required to establish the same pressure drop in all channels. Usually an appropriate flow resistance (larger than by the bed of catalyst particles is put in front of the reactor channel (see Figures 2.5a and 2.5b).

Figure 2.5a. Reactor for high-throughput testing of catalytic materials at temperatures up to about 1,200 °C (e.g. for HCN formation from methane and ammonia using capillaries to compensate for equal pressure drop in all reactor channels (Moehmel *et al.*, 2008).

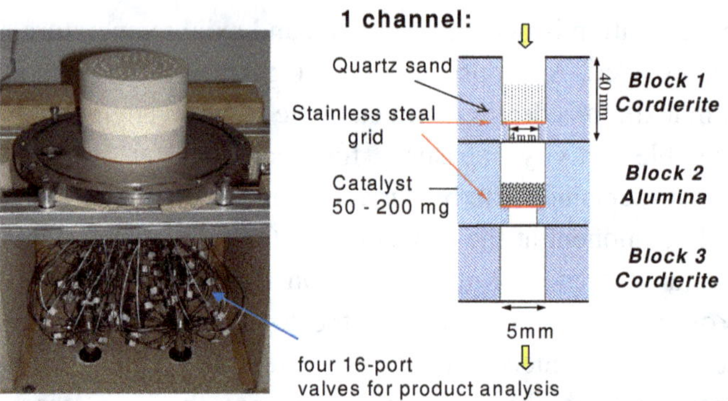

Figure 2.5b. Reactor (insulation not shown) for high-throughput testing of catalytic materials (e.g. catalysts for the oxidative dehydrogenation of alkanes at temperatures (up to about 500 °C) using quartz sand to compensate for equal pressure drop in all reactor channels.

Finally, a multifold single-bead reactor is mentioned (see Figure 2.6), in which only one catalyst bead is hosted in each single channel. The flexible concept of this reactor type allows the adaptation of several techniques for analysis of the product gas leaving the channels. Analysis can take place either using sequential techniques that retrieve gas samples from single catalytic beads, typically via a capillary, or applying integral techniques monitoring all samples at one specified time. MS, GC/MS, GC, or dispersive or non-dispersive IR can perform fast sequential analysis. It is also possible to apply integral analysis techniques that allow true parallel analysis — for example — IR-thermographic or photoacoustic spectroscopy (Schunk *et al.*, 2004).

Frequently, in evaluating catalyst performance emphasis is put on the selectivity of a desired product and not only on activity. In such cases account should be taken of the fact that selectivity depends on the degree of conversion of the key feed molecules. Under certain circumstances, e.g. selective oxidation of hydrocarbons, the flow conditions can be adjusted in such a way that the degree of oxygen conversion is 1; thus, the selectivity of all the catalysts can be compared. Otherwise it may be advisable to vary the total flow-rate to the multi-channel reactor; in this way, selectivity can be obtained as a function of the degree of conversion. Comparable selectivity data for the different catalysts can be

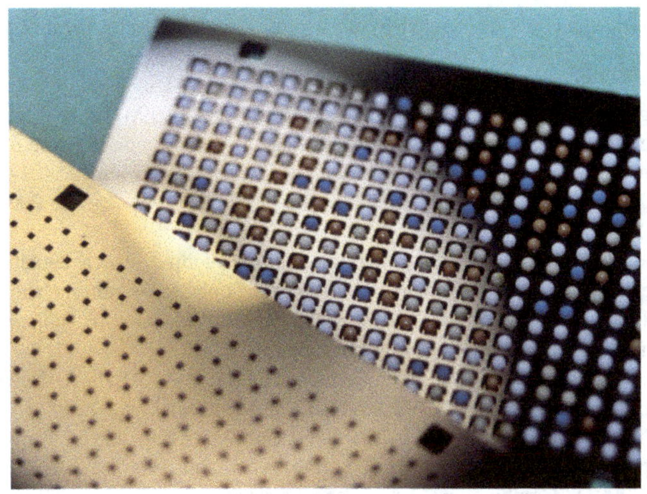

Figure 2.6. Single-bead reactor (625 parallel channels) for testing of catalysts in gas-phase reactions. (Photograph courtesy of hte Aktiengesellschaft Heidelberg)

estimated by interpolating the catalyst-specific dependency for a given degree of conversion; extrapolation is less to be recommended.

Note of caution: In all catalytic testing experiments the absence of any external and internal transport limitations within the catalyst particle and at its outer surface has to be ascertained; otherwise, results regarding activity and selectivity, by which catalytic performance is to be characterized, may not be of any significance.

Bibliography

Buyevskaya, O.V., Wolf, D. and Baerns, M. (2000). Ethylene and propene by oxidative dehydrogenation of ethane and propane: Performance of rare-earth oxide-based catalysts and development of redox-type catalytic materials by combinatorial methods. *Catal. Today*, 62, 1, 91–99.

Cong, P., Doolen, R.D., Fan, Q. Giaquints, S., Guan, R.W., Farland, E.W., Pooraj, D.M., Self, K., Turner, H.W. and Weinberg, W.H. (1999). High-throughput synthesis and screening of combinatorial heterogeneous catalyst libraries. *Angew. Chem. Int. Ed.*, 38, 484.

Grubert, G., Kondratenko, E., Kolf, S., Baerns, M., van Geem, P. and Parton, R. (2003). Fundamental insights into the oxidative dehydrogenation of ethane to ethylene over catalytic materials discovered by an evolutionary approach, *Catal. Today*, 81, 3, 337–345.

Moehmel, S., Steinfeldt, N., Engelschalt, S., Holena, M., Kolf, S., Baerns, M., Dingerdissen, U., Wolf, D., Weber, R. and Bewersdorf, M. (2008). New catalytic materials for the high-temperature synthesis of hydrocyanic acid from methane and ammonia by high-throughput approach. *Appl. Catal. A: General* 334, 1–2, 73–83.

Newsam, J.M., Schunk, S.A. and Klein, J. DE 100059890 A1; 1 Dec. 2000 and WO 02/43860 A2; 30 Nov. 2001.

Rodemerck, U. and Baerns, M. (2004). High-throughput experimentation in the development of heterogeneous catalysts — Tools for synthesis and testing of catalytic materials, in: Baerns M. (ed.), *Basic Principles in Applied Catalysis,* Springer, Berlin Heidelberg New York, pp. 261–279.

Schunk, S.A., Demuth, D., Cross, A., Gerlach, O., Haas, A., Klein, J., Newsam, J.M., Sundermann, A., Stichert, W., Strehlau, W., Vietze, U. and Zech, T. (2004). In: *High-Throughput Screening in Chemical Catalysis* (Hagemeyer, A., Strasser, P., Volpe, jr., A.V. (eds.) Wiley-VCH Verlag, Weinheim, 19–61.

Senkan, S. (2001). Combinatorial heterogeneous catalysis — A new path in an old field. *Angew. Chem. Int. Ed.* 40, 2, 312–329.

Chapter 3

Mathematical Methods of Searching for Optimal Catalytic Materials

3.1. Introduction

Many properties and conditions influence the capability of materials to serve as catalysts for a particular reaction. First, the composition of the material is important, identified by the chemical elements or compounds present, and by their respective mass or molar fractions in the material. Furthermore, the structure and texture of the material, which are determined mainly by the preparation method employed, play an important role. The composition and structure of the bulk and the surface are interconnected with further properties of the material, such as acidity, basicity, redox potential, and electronic conductivity; these properties are usually denoted as *descriptors* (cf. Klanner, 2004; Klanner *et al.*, 2004; Farrusseng *et al.*, 2005). Finally, the catalytic performance of a material does not depend solely on the material itself, but also on the conditions to which that material is exposed in the reaction, at a particular temperature, partial pressures of the reactants, total pressure, and space velocity of the feed. The descriptors of the material and the reaction conditions together are called *input variables* in the sequel.

In a catalytic experiment, a certain number of catalytic materials is tested, chosen from a very comprehensive set of potential materials. The number of materials to be examined is affected by the following aspects:

(i) *Ranges* of individual input variables.
(ii) Whether a particular input variable can have *only particular prespecified values*, or whether it can assume *any value* within the

corresponding range, restricted only through the finite discernibility due to experimental error.

(iii) *Experimental error* of a particular input variable. Since some experimental error for a continuous variable always exists, the number of discernible values within its range is always finite, although this number may be quite high. For example, a range of fractions 0–20% and an experimental error of 0.1% entail 201 discernible values.

(iv) *Constraints* on input variables, such as the constraint that the fractions of all components should sum up to 100%, or constraints on the number of non-zero fractions. The latter actually express constraints on the number of components in a catalytic material, for example, 2–3 active components + 1 dopant.

3.2. Statistical Design of Experiments

The task of finding, for a given set of potential catalytic materials, a subset of *representatives conveying the required information about the whole set* has for nearly a century been addressed by methods of *statistical design of experiments* (DOE). A very simple kind called *factorial design* is used if the required information represents the impact of any combination of possible values of some $\boldsymbol{n}$ independent factors. If for $\boldsymbol{i} = 1,\ldots,\boldsymbol{n}$ the $\boldsymbol{i}$-th factor can assume f_i different values, the factorial design requires $f_1\cdot\ldots\cdot f_n$ representatives. For example, let a catalytic material have a fixed fraction of support chosen from two possibilities, and active components chosen from three possibilities, among which one is always present, whereas each of the remaining two is either absent or its ratio to the first one can assume one of three given values, and can have one dopant which either is absent or assumes a given value. Then $\boldsymbol{n} = 4$ independent factors exist, $f_{\text{support}} = 2$, $f_{\text{2nd active componetnt}} = f_{\text{3rd active componetnt}} = 4$, $f_{\text{dopant}} = 2$, and the factorial design needs

$$f_{\text{support}} \cdot f_{\text{2nd active componetnt}} \cdot f_{\text{3rd active componetnt}} \cdot f_{\text{dopant}} = 64 \text{ representatives.}$$

On the other hand, assuming that there is no interaction between different factors and one requires only information about the impact of any single factor in isolation, one requires only $n + 1$ representatives. In the above example, $n + 1 = 5$.

In both the above methods, the number of representatives can vary from experiment to experiment, depending on the number of independent factors, and in factorial design also on the number of their possible values. Since 1970s, *computer-aided statistical DOE methods* have been developed, which deal with a more ambitious and computationally much more demanding task: to choose for a given set of possible materials a *representative subset of particular size such that the amount of required information about the whole set is maximal among all subsets of that size.*

Among computer-aided DOE methods, the *D-optimal design* is the most frequently used. It relies on a model that assumes a dependent variable, such as yield, degree of conversion or selectivity, which linearly depends on input variables and interactions between them. The information maximised in the D-optimal design is an information measure called *Fisher information* of that model; more precisely, it is the determinant of the matrix of that information measure (actually, the term "D-optimal" is due to the fact that the determinant is maximised). Moreover, from statistics it is known that maximising that information is equivalent to minimising the volume of the confidence ellipsoid for parameter estimates in the underlying linear model. To illustrate these principles again using an example, a catalytic material is considered that consists of a support material and four active components. For this material, it is assumed that the yield y_i of the reaction catalyzed with the i-th catalytic material from the representative subset depends on the fraction $x_{i,0}$ of the support and the fractions $x_{i,1}$, $x_{i,2}$, $x_{i,3}$ of the active components 1, 2 and 3 in this material. The fraction $x_{i,4}$ of the active component 4 is obtained from the condition $x_{i,0} + x_{i,1} + x_{i,2} + x_{i,3} + x_{i,4} = 1$. Moreover, the yield y_i is also assumed to depend on interactions between any two of the fractions $x_{i,1}$, $x_{i,2}$, $x_{i,3}$ and on interactions between $x_{i,0}$ and any pair of those three active-component fractions. Rewritten in a formula, these assumptions read:

$$\begin{aligned} y_i = {} & \beta_0 x_{i,0} + \beta_1 x_{i,1} + \beta_2 x_{i,2} + \beta_3 x_{i,3} + \beta_{01} x_{i,0} x_{i,1} + \beta_{02} x_{i,0} x_{i,2} + \beta_{03} x_{i,0} x_{i,3} \\ & + \beta_{12} x_{i,1} x_{i,2} + \beta_{13} x_{i,1} x_{i,3} + \beta_{23} x_{i,2} x_{i,3} + \beta_{012} x_{i,0} x_{i,1} x_{i,2} + \beta_{013} x_{i,0} x_{i,1} x_{i,3} \\ & + \beta_{023} x_{i,0} x_{i,2} x_{i,3}, \end{aligned}$$

where $\beta_0, \beta_1, \ldots \beta_{023}$ denote the unknown model parameters. The information, which is maximised in the D-optimal design, is then the determinant det($\boldsymbol{R}'\boldsymbol{R}$) with $\boldsymbol{R}$ denoting the 13-column matrix with rows

$$(\boldsymbol{x}_{i,0}\ \boldsymbol{x}_{i,1}\ \boldsymbol{x}_{i,2}\ \boldsymbol{x}_{i,3}\ \boldsymbol{x}_{i,0}\boldsymbol{x}_{i,1}\ \ldots\ \boldsymbol{x}_{i,2}\boldsymbol{x}_{i,3}\ \boldsymbol{x}_{i,0}\boldsymbol{x}_{i,1}\boldsymbol{x}_{i,2}\ \boldsymbol{x}_{i,0}\boldsymbol{x}_{i,1}\boldsymbol{x}_{i,3}\ \boldsymbol{x}_{i,0}\boldsymbol{x}_{i,2}\boldsymbol{x}_{i,3})$$

and $\boldsymbol{R}'$ standing for the transpose of $\boldsymbol{R}$. As was recalled above, maximising that information is equivalent to minimising the volume of the confidence ellipsoid for the estimates of $\beta_0, \beta_1, \ldots \beta_{023}$.

For a comprehensive explanation of various statistical DOE methods, competent monographs are available — both general ones (Box *et al.*, 1978; Atkinson and Donev, 1992; Funkenbusch, 2005), and those addressing specifically chemical applications (Carlson, 2005; Deming and Morgan, 2005). Examples of such applications in catalytic research can be found in Neele *et al.* (1999), Ramos *et al.* (2000), Corma *et al.* (2003), Tagliabue *et al.* (2003), Bricker *et al.* (2004), and Hendershot *et al.* (2004).

3.3. Optimisation Methods for Empirical Objective Functions

Different statistical methods for experiment design use different criteria for choosing which material to test experimentally from among all those considered possible for testing; however, they always apply the respective criteria uniformly to the whole set of possible materials. This could be quite impractical if the most interesting catalytic materials are not uniformly distributed in the space of possible materials, but instead form only one or several small clusters. This happens in the frequently occurring event that interest is only in catalytic materials with sufficiently high performance (in terms of yield, degree of conversion, selectivity, etc.). Indeed, high values of such performance measures are typically achieved only in small areas of contiguous compositions (see Figure 3.1).

In such situations it is more relevant to use a method designed specifically to seek, for a given objective function, locations at which the function values are maximised. Since each such method equivalently seeks locations in which the negative of the objective function takes its minimal value, the more general term *optima of the objective function*

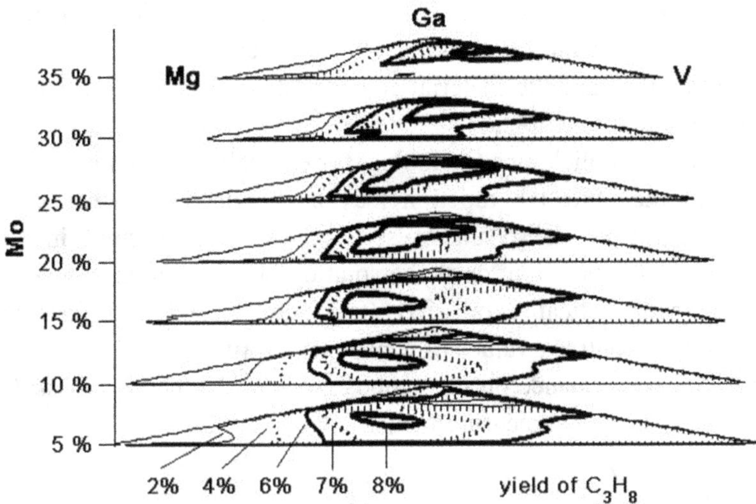

Figure 3.1. Example of a dependency of yield on catalyst composition. This example shows the dependency of propene yield on the fractions of Mg, V, Ga and Mo in the oxidative dehydrogenation of propane. The approximation was obtained by means of artificial neural networks on data analysed in Cawse *et al.* (2004).

is used instead of *maxima* and *minima.* The process of searching for such optima by means of a corresponding algorithm is called *function optimisation*, and the computer-aided methods involved are called *optimisation methods.* Before the overview of methods of this type is undertaken, it will be appropriate to make two introductory remarks:

1. Whether a location is an optimum of the objective function in the space of input variables depends on other locations with which it is compared. Sometimes the function value in a location is not surpassed by values in locations within a certain neighborhood, but outside that neighborhood there are locations with a higher function value. For example, varying the fraction of individual elements in the catalytic material within a certain small range does not lead to a material with a better performance, but varying some of those fractions outside that range can lead to such a performance. In such a case, the material is said to have a *locally optimal performance*, or in mathematical terms, to be a *local optimum* of the function describing

the dependency of its performance on the composition. On the other hand, if even varying any of the fractions within the whole range of their admissible values does not lead to a better performance, then it is said that the catalytic material considered corresponds to a *global optimum* of that function.

2. In most other applications (including, for example, reaction kinetics), values of the objective function may be obtained analytically — that is, either as the result of setting the function input into a mathematical expression, or as the solution of an equation described with a mathematical expression (for example, of a differential equation). In contrast, values of functions describing the dependence of catalyst performance on its composition are obtained *empirically,* through experimental measurements.

3.4. Evolutionary Optimisation: The Main Approach to Seek Optimal Catalysts

Evolutionary optimisation methods are *stochastic methods.* This means that the available information about the objective function is complemented by random influences. The term "evolutionary" refers to the fact that the way of incorporating random influences into the optimisation process has in those methods been inspired by *biological evolution.* The most frequently used and most highly elaborated representatives of evolutionary methods are *genetic algorithms* (GAs), explained below, in which the incorporated random influences attempt to mimic the *evolution of a genotype*. Basically, this method involves:

- Randomly exchanging coordinates between two particular locations in the input space of the objective function (*recombination, crossover*).
- Randomly modifying coordinates of a particular location in the input space of the objective function (*mutation*).
- *Selecting* the locations for crossover and mutation (parent locations) according to a probability distribution either uniform or skewed towards locations at which the objective function takes high values (the latter being a probabilistic expression of the survival-of-the-fittest principle).

Detailed treatment of various kinds of genetic algorithms, as well as of other traditional evolutionary optimisation methods, can be found in specialised monographs (Goldberg, 1989; Bäck, 1996; Mitchell, 1996; Fogel, 1999; Vose, 1999; Wong and Leung, 2000; Freitas, 2002; Reeves and Rowe, 2003; Bartz-Beielstein, 2006; Bandyopadhay and Pal, 2007; Schaefer, 2007). In addition, monographs from the 2000s introduce novel kinds of evolutionary optimisation which have not yet been encountered in catalytic applications but are likely to be employed there in the future; in particular, the differential evolution approach (Feoktistov, 2006), and the estimation of distribution algorithms (Larranaga and Lozano, 2002). In this chapter, only those features of GAs will be highlighted that are particularly important for the development of catalytic materials:

1. The meaning of the individual coordinates of locations in the input space of the objective function is strongly problem-dependent. In catalyst design, the coordinates typically convey some of the following meanings:
   (i) *Qualitative composition* of the catalytic material — that is, of which active components it consists, whether it contains dopants and if so, which ones, whether it is supported, and what its support is.
   (ii) *Quantitative composition* of the catalytic material, that is, the fractions of the various components mentioned in (i).
   (iii) *Preparation of the catalytic material*, its individual steps and their quantitative characterisations, such as temperatures or times.
   (iv) *Reaction conditions* of the catalyzed reaction.

   There is an intimate connection between qualitative and quantitative composition of catalytic materials. The presence of a particular component in the catalytic material is equivalent to the fraction of that component being non-zero. This has consequences for the employed evolutionary algorithm: it has to guarantee that this equivalence cannot become invalidated through its operations. For example, if the presence of a particular component in one of the parent materials and the absence of that component in the other parent material are exchanged during recombination, then the fraction of that component also has to be changed at the same time. Similarly, if a mutation eliminates a certain component from the catalytic material, the fraction of that component must be simultaneously set to zero. In this context, it is useful to differentiate between *quantitative mutation*, which modifies merely the quantitative composition of the catalytic material, without affecting its qualitative composition, and *qualitative mutation*, which also modifies its qualitative composition (see Figure 3.2).

2. Crossover and mutation operations can be applied to many individuals simultaneously; therefore, the genetic algorithm can *follow many optimisation paths in parallel.* Moreover, optimisation proceeds between subsequent iterations for different paths independently. Because of the biological inspiration of genetic algorithms, individual iterations of a genetic algorithm are called *generations,* and all locations in which the value of the objective function is considered in a particular generation (for example, all catalytic materials of which the performance has been measured in

Crossover

B = 0	Fe = 0	Ga = 40	La = 0	**Mg = 34**	**Mn = 0**	**Mo = 11**	V = 15
				+			
B = 0	Fe = 0	Ga = 32	La = 0	**Mg = 17**	**Mn = 18**	**Mo = 10**	V = 23
				↓			
B = 0	Fe = 0	Ga = 40	La = 0	**Mg = 17**	**Mn = 18**	**Mo = 10**	V = 15
				+			
B = 0	Fe = 0	Ga = 32	La = 0	**Mg = 34**	**Mn = 0**	**Mo = 11**	V = 23

Qualitative mutation

B = 0	Fe = 0	Ga = 32	La = 0	Mg = 17	**Mn = 18**	Mo = 10	V = 23
				↓			
B = 0	Fe = 0	Ga = 49	La = 0	Mg = 21	**Mn = 0**	Mo = 12	V = 18

Quantitative mutation

B = 0	Fe = 0	**Ga = 32**	La = 0	**Mg = 17**	**Mn = 18**	**Mo = 10**	**V = 23**
				↓			
B = 0	Fe = 0	**Ga = 48**	La = 0	**Mg = 15**	**Mn = 1**	**Mo = 16**	**V = 20**

Figure 3.2. Illustration of operations used in genetic algorithms; the values in the examples are mass fractions of active elements in the catalytic material expressed in mol-%.

that generation) are called a *population*. The fact that GAs follow many optimisation paths in parallel is actually the main reason for their attractiveness in high-throughput catalyst development because a straightforward correspondence can be established between those optimisation paths and channels of the high-throughput reactor in which the materials are experimentally tested.

3. Since GAs do not use derivatives, they are not attracted to local optima. Moreover, the random variables incorporated into recombination, mutation and selection enable the optimisation paths to leave the attraction area of the nearest local optimum, and to continue searching for a global one.
4. The optimisation of catalytic materials is always a *constrained optimisation*, either continuous, or mixed (i.e., with constraints involving both continuous and discrete variables), or finally, mixed optimisation with constraints involving only continuous variables. That important topics will be treated in a separate subsection below.

Due to the problem-dependency of GA inputs, for solving optimisation problems in specific application areas such in the development of catalytic materials, it is quite difficult to use general GA software, such as Matlab's Genetic Algorithm and Direct Search Toolbox (2004). Indeed, such general GA software only optimises functions with input spaces of low-level data types, such as vectors of real numbers and bit-strings. And encoding the qualitative and quantitative compositions of catalytic materials, and their preparation and reaction conditions with low-level data types is a tedious and error-prone activity. Furthermore, it requires a great deal of mathematical erudition.

For all these reasons, it is not surprising that — apart from early attempts to use general GA software to optimise the distribution of active sites of two catalytic components (McLeod *et al.*, 1997; McLeod and Gladden, 2000), and of rarely encountered multiobjective optimisation (Gobin *et al.*, 2007; Gobin and Schüth, 2008) — the application of GAs in this area did not take the route of using general GA software, but instead that of developing specific algorithms for the optimisation of catalytic materials. The implementation of such an algorithm for solid catalyst development was first undertaken by D. Wolf and colleagues at the Leibniz Insittute for Catalysis (LIKAT) in Germany, in the late 1990s (Buyevskaya *et al.*, 2000; Wolf *et al.*, 2000; Buyevskaya *et al.*, 2001; Rodemerck *et al.*, 2001), followed by a modified version in 2001 (Grubert *et al.*, 2003; Duff *et al.*, 2004; Rodemerck *et al.*, 2004; Grubert *et al.*, 2006); similar algorithms were later also developed at other institutions (Corma *et al.*, 2003; Serra *et al.*, 2003; Kirsten and Maier, 2004; Omata *et al.*, 2004; Serra and Corma, 2004; Watanabe *et al.*, 2004; Clerc *et al.*, 2005; Corma and Serra, 2005; Ohrenberg *et al.*, 2005; Pereira *et al.*, 2005; Serra *et al.*, 2007). Experience gained with these algorithms shows that any new method of catalyst design can substantially decrease the usefulness of an earlier-implemented specific GA after several years. To retain a high usefulness of the implementation for a long period requires, during the development of that implementation, the anticipation of all catalytic materials and their preparation methods for which the implemented algorithm might need to be employed in the future. Nevertheless, there is no guarantee that all of them really can be anticipated during development. An approach that avoids this difficulty, namely generating problem-tailored genetic algorithms automatically at the time when they are needed, will be described in the next chapter.

3.4.1. *Dealing with Constraints in Genetic Optimisation*

In the context of constraint optimisation, it is important that genetic algorithms have been originally developed as an unconstrained optimisation method. Nevertheless, repeated attempts have been made to modify them for constraints. Basically, these attempts belong to one of (or combine several of) the following approaches (Coello, 2002; Fonseca and Fleming, 1995; Michalewicz and Schonauer, 1996; Michalewicz and Schmidt, 2003; Reid, 1996; Chu and Beasley, 1998; Fung *et al.*, 2002; Bunnag and Sun, 2005; Li *et al.*, 2005):

(i) *Ignoring* offsprings that are not feasible with respect to constraints and not including them in the new population. Because in constrained optimisation the global optimum frequently lies on a constraint boundary, ignoring unfeasible offsprings may lead to discarding information even on offsprings that are very close to the optimum. Moreover, for some genetic algorithms this approach can lead to a deadlock situation where the algorithm cannot find a whole population of feasible offsprings.

(ii) To modify the objective function by superposing a *penalty for unfeasibility*. Such a penalty can be constant (the offspring is penalised for being unfeasible), or it can reflect either the amount of constraint violation or the effort needed to correct the unfeasibility. This approach works well if theoretical considerations allow an appropriate choice of penalty function. If, on the other hand, some heuristic penalty function is used, then its values typically turn out to be either too low, which can allow the optimisation paths to remain permanently outside the feasibility area, or too high, thus suppressing information on the value of the objective function for unfeasible offsprings.

(iii) *Repairing the original unfeasible offspring* through modifying it so that all constraints become fulfilled. Some methods simultaneously repair all constraints not fulfilled by the original offspring, others construct a sequence of modifications in which the subsequent modification always fulfils one more constraint than the previous one. This approach can once again lead to discarding information on some offsprings close to the global optimum, so various modifications of the repair approach have been proposed that dispense with full feasibility but preserve some information on the original offsprings, e.g., repairing only randomly selected offsprings (with a prescribed probability), or using the original offspring but with the value of the objective function of its repaired version. In the context of evolutionary optimisation of catalytic materials, such an approach was recently used in Gobin *et al.* (2007), and Gobin and Schüth (2008).

(iv) Adding feasibility/unfeasibility as *another objective function*, thus transforming the original constrained optimisation task into a task of optimisation with respect to multiple goals. Unfortunately, this new task similarly remains far from the original purpose of genetic algorithms and other evolutionary methods, and is similarly difficult for solving with them as the original one.

(v) *Modifying the recombination and/or mutation operator* in such a way that it becomes closed with respect to the set of feasible solutions. Hence, it is the case that also a mutation of a point fulfilling all the constraints or a recombination of two such points has to fulfil them. Unfortunately, such a modification always requires sufficient knowledge about the considered constraints, so this approach cannot be elaborated in a general setting. On the other hand, this is the most usual way of dealing with constraints in the genetic algorithms used in catalysis. An example will be given in the next chapter.

Experience with those approaches indicates that each has its own specific problems, and no single approach provides an ultimate way of dealing with constraints that is generally better than all the others.

3.5. Other Stochastic Optimisation Methods

There are several *other stochastic methods* for optimisation of an objective function, for example, simulated annealing, stochastic tabu search, multilevel single linkage, topographical optimisation, stochastic hill-climbing, stochastic tunnelling, and stochastic branch-and-bound (Otten and Ginneken, 1989; Zabinski, 2003). Like GAs, they typically only compare function values at different locations. On the other hand, function values alone contain but little information; therefore, the search for the optimum progresses very slowly.

Simulated annealing has been sometimes used in the design of solid catalysts, although less frequently than GAs (Li *et al.*, 1999; McLeod and Gladden, 2000; Eftaxias *et al.*, 2001; Holzwarth *et al.*, 2001; Senkan, 2001). Basically, the way in which the simulated annealing method searches for the global minimum of an objective function mimics the way in which a crystalline substance reaches the ground state with minimal energy in a process of heating and then slow cooling. The method works by iteratively proposing changes and either accepting or rejecting each of them. There are various criteria for the acceptance or rejection of proposed changes; the most frequently encountered criterion is called *metropolis*. This accepts the change unconditionally only if the objective function decreases; otherwise, it accepts the change with an *exponentially distributed probability* that depends on a parameter being called *temperature*, which is equal to the mean of that exponential distribution. Thus, if the temperature is very high, nearly all changes are accepted and the method simply moves the system through various states, irrespective of the values of the objective function. Consequently, if the method starts at a higher temperature which is then gradually decreased, it brings the system to a state with a minimum of the objective function, but still allowing it to escape that state if the minimum is not global (hence, the temperature is not zero). In particular, when searching for catalyst materials with high performance, the simulated annealing method can leave a locally optimal catalyst, and continue with lower-performance materials, in order to ultimately find a globally optimal catalyst.

During the last decade two novel stochastic optimisation methods came into use which, like evolutionary algorithms, are based on heuristics inspired by nature. These methods are particle swarm optimisation (Kennedy and Eberhart, 2001), and ant colony optimisation (Dorigo and Stützle, 2004).

3.6. Deterministic Optimisation

Optimisation methods that rely solely on the available information about the response surface of the optimised function, with no randomness involved, are referred to as *deterministic*. From the overall perspective of function optimisation, the wide variety of such methods means that they are more frequently used than stochastic methods (Floudas, 2000; Snyman, 2005). In such methods, the same starting location of the optimisation procedure always leads to iterating through the same locations in the input space of the objective functions. Before we discuss the role of deterministic methods in the optimisation of catalytic materials, it should be recalled that according to the information they use, all such methods can be divided into three large groups:

1. *Methods using only information about function values*. As was already mentioned in the context of stochastic optimisation, such

methods tend to find a global optimum rather than a local one (contrary to methods of the groups 2 and 3 below), but they are very slow. The most frequently encountered example is the *simplex method*. This produces a sequence of simplices, that is, of polyhedra with $\boldsymbol{n} + 1$ vertices, where $\boldsymbol{n}$ refers to the dimension of the input space of the objective function. The vertices of the first simplex of the sequence are typically chosen at random, and any further simplex is obtained from the previous one by replacing the vertex in which the value of the objective function was the worst (that is, the lowest value in the case of maximisation, and the highest value in the case of minimisation). Other methods belonging to this group, used in the design of catalytic materials, are the *holographic research strategy* and the *sequential weight- increasing factor technique*.

2. *Methods using, in addition to function values, information about first partial derivatives, that is, about the gradient*. The gradient of a function has the property that its direction coincides with the direction of the fastest increase of function values in a neighbourhood of the considered point and is opposite to the direction of their fastest decrease. Thus, if the optimisation path follows, from some location in the input space, the direction of the gradient of the objective function, then the function value increases with the highest possible speed along that path (at least, in the immediate neighbourhood of that location). Similarly, if the path follows the direction opposite to the gradient, then the function value decreases with the highest possible speed. However, following the direction of the gradient (respectively, opposite to the gradient) of the objective function does not in general allow the optimisation path to reach its global maximum (respectively, minimum), but only a local one. More precisely, for every maximum, global as well as local, there exists an area that an optimisation path will never leave if it starts within that area and follows in each location the direction of the gradient of the objective function. The area is called *attraction area* of the considered maximum and the maximum is said to be an *attractor* of that area. Global and local minima are also attractors, although the optimisation paths in their attraction areas follow in each location the direction opposite to the gradient. Simple examples of methods of this

group are different variants of *steepest descent*, which directly employs the property of gradients above recalled, and searches for a minimum of the function in such a way that the next location in a sequence of iterations is chosen in the direction opposite to the gradient in the current location. Individual variants of this method differ in terms of how they choose the next location in that direction. More sophisticated representatives of this group include several *methods of conjugate gradients*.

3. *Methods using in addition to function values information about partial derivatives up to the second order*. Like gradient-based methods, such methods search only for local maxima or minima of the objective function, according to the attraction area in which the optimisation path starts. However, second-order derivatives allow one to construct a quadratic approximation of the function, whereas the gradient approximates the function only linearly. Close to a maximum or minimum, a quadratic approximation of a function is more accurate than a linear one; therefore, methods belonging to this group can localise the optima faster, once they get to their proximity. On the other hand, linear approximations are frequently more accurate far from any optima. Therefore, methods of this group are most often used as combined methods, which switch between the behavior of second-order methods close to an optimum and the behavior of a gradient-based method far from the optimum. The most frequently encountered method of that kind is the *Levenberg-Marquardt method*, whereas using only second-order derivatives leads in the most simple case to the *Gauss-Newton method*.

3.6.1. *Utilizability of Methods with Derivatives in Catalysis*

For optimisation of empirical functions in the development of solid catalytic materials, it is unfortunately impossible to employ methods using partial derivatives, that is — methods from groups 2 and 3. The reasons for this impossibility are connected with the *character of catalytic experiments*, and can be summarised as set out below:

Since a *mathematical expression* for the dependency of performance of the catalytic material on the various input variables is not known, mathematical expressions for its partial derivatives cannot be obtained either.

To obtain sufficiently accurate *numerical estimates* of the partial derivatives, small differences between values of the dependent variable need to be recorded; typically, differences should not be larger than 0.1% of the function value in the respective location. However, for dependent variables occurring in catalyst optimisation (such as yield, degree of conversion, selectivity), such differences commonly lie within the *experimental error*.

Even if the experimental setting led to a lower experimental error (or if sufficient accuracy of the estimates were not required), obtaining numerical estimates of the gradient must be given up for practical reasons. Indeed, it is known from numerical mathematics that to obtain a numerical estimate of the gradient in any location of an $\boldsymbol{n}$-dimensional input space of the objective function would require one to empirically evaluate the function in at least ($\boldsymbol{n}$ + 1) locations located *very close to each other*. For example, imagine that the objective function describes the dependence of yield on the composition of catalytic material with components from a pool of 15 elements. In this case, numerically estimating its gradient in one single location would require one to test 16 materials with almost the same composition. This is clearly not affordable for cost and time reasons.

That is why in the search for solid catalytic materials, only deterministic optimisation methods from group 1 have occasionally been used (Holzwarth *et al.*, 2001; Huang *et al.*, 2003; Végvári *et al.* 2003; Lin *et al.*, 2005; Tompos *et al.*, 2005; Tompos *et al.*, 2007). On the other hand, in the context of high-throughput development of solid catalysts, it was pointed out that the best-known representative of such methods — the simplex method — allows an easy adaptation to produce any prescribed number of sequences of simplices in parallel (Holzwarth *et al.*, 2001). Consequently, the method is able to propose in each step a *prescribed number of new catalytic materials in parallel*, which provides a straightforward correspondence with their subsequent testing in a multichannel reactor, as in the case of evolutionary algorithms.

Bibliography

Atkinson, A.C. and Donev, A.N. (1992). *Optimum Experimental Designs*, Oxford University Press, Oxford, 328 p.

Bandyopahay, S. and Pal, K. (2007). *Classification and Learning Using Genetic Algorithms*, Springer, Berlin, 311 p.

Bäck, T. (1996). *Evolutionary Algorithms in Theory and Practice: Evolution Strategies, Evolutionary Programming, Genetic Algorithms*, Oxford University Press, New York, 314 p.

Bartz-Beielstein, T. (2006). *Experimental Research in Evolutionary Computation*, Springer, Berlin, 214 p.

Box, E.P., Hunter, W.G. and Hunter, J.S. (1978). *Statistics for Experimenters: An Introduction to Design, Data Analysis, and Model Building*, Wiley, New York, 653 p.

Bricker, M.L., Sachtler, J.W.A., Gillespie, R.D., McGonegal, C.P., Vega, H., Bem, D.S. and Holmgren, J.S. (2004). Strategies and applications of combinatorial methods and high throughput screening to the discovery of non-noble metal catalyst, *Appl. Surf. Sci.*, 223, 109–117.

Bunnag, D. and Sun, M. (2005). Genetic algorithms for constrained global optimization in continuous variables, *Appl. Math. Comput.*, 171, 604–636.

Buyevskaya, O.V., Wolf, D. and Baerns, M. (2000). Ethylene and propene by oxidative dehydrogenation of ethane and propane: Performance of rare-earth oxide-based catalysts and development of redox-type catalytic materials by combinatorial methods, *Catal. Today*, 62, 91–99.

Buyevskaya, O.V., Bruckner, A., Kondratenko, E.V., Wolf, D. and Baerns, M. (2001). Fundamental and combinatorial approaches in the search for and optimization of catalytic materials for the oxidative dehydrogenation of propane to propene, *Catal. Today*, 67, 369–378.

Carlson, R. (2005). *Design and Optimisation in Organic Synthesis*, 2nd Ed., Elsevier, Amsterdam, 596 p.

Cawse, J.N., Baerns, M. and Holeňa, M. (2004). Efficient discovery of nonlinear dependencies in a combinatorial data set, *J. Chem. Inf. Comput. Sci.*, 44, 143–146.

Chu, P.C. and Beasley, J.E. (1998). Constraint handling in genetic algorithms: The set partitioning problem, *J. Heuristics*, *4*, 323–358.

Clerc, F., Lengliz, M., Farrusseng, D., Mirodatos, C., Pereira, S.R.M. and Rakotomata, R. (2005). Library design using genetic algorithms for catalysts discovery and optimization, *Rev. Sci. Instrum.*, 76, 062208.

Coello, C.A. (2002). Theoretical and numerical constraint-handling techniques used with evolutionary algorithms: A survey of the state of the art, *Comput. Methods Appl. Mech. Eng.*, 191, 1245–1287.

Corma, A., Serra, J.M. and Chica, A. (2003). Discovery of new paraffin isomerization catalysts based on SO_4^{2-}/ZrO_2 and WO_x/ZrO_2 applying combinatorial techniques, *Catal. Today*, 81, 495–506.

Corma, A. and Serra, J.M. (2005). Heterogeneous combinatorial catalysis applied to oil refining, petrochemistry and fine chemistry, *Catal. Today*, 107–108, 3–11.

Deming, S.N. and Morgan, S.L. (2005). *Experimental Designs: A Chemometric Approach*, 2nd Ed., Elsevier, Amsterdam, 454 p.

Dorigo, M. and Stützle, T. (2004). *Ant Colony Optimization,* MIT Press, Cambridge (MA), 319 p.

Duff, D.G., Ohrenberg, A., Voelkening, S. and Boll, M. (2004). A screening workflow for synthesis and testing of 10,000 heterogeneous catalysts per day - Lessons learned, *Macromol. Rapid Commun.*, 25, 169-177.

Eftaxias, A., Font, J., Fortuny, A., Giralt, J., Fabregat, A. and Stüber. F. (2001). Kinetic modelling of catalytic wet air oxidation of phenol by simulated annealing, *Appl. Catal., B: Environ.*, 33, 175–190.

Farrusseng, D., Klanner, C., Baumes, L., Lengliz, M., Mirodatos, C. and Schüth, F. (2005). Design of discovery libraries for solids based on QSAR models, *QSAR Comb. Sci.*, 24, 78–93.

Feoktistov, V. (2005). *Differential Evolution in Search of Solutions*, Springer, Berlin, 195 p.

Floudas, C.A. (2000). *Deterministic Global Optimization: Theory, Methods and Applications*, Kluwer, Dordrecht, 739 p.

Fogel, D.B. (1999). *Evolutionary Computation: Toward a New Philosophy of Machine Intelligence,* 2nd Ed., IEEE Press, New York, 270 p.

Fonseca, C.M. and Fleming, P.J. (1995). An overview of evolutionary algorithms in multiobjective optimization, *Evol. Comput.*, 3, 1–16.

Freitas, A.A. (2002). *Data Mining and Knowledge Discovery with Evolutionary Algorithms,* Springer, Berlin, 264 p.

Funkenbusch, P.D. (2005). *Practical Guide to Designed Experiments: A Unified Modular Approach*, Dekker, New York, 197 p.

Fung, R., Tang, J. and Wang, D. (2002). Extension of a hybrid genetic algorithm for nonlinear programming problems with equality and inequality constraints, *Comput. Oper. Res.*, 29, 261–274.

Genetic Algorithm and Direct Search Toolbox (2004). The MathWorks, Inc., Natick, 268 p.

Gobin, O.C., Martinez, J.A. and Schüth, F. (2007). Multi-objective optimization in catalytical chemistry applied to the selective catalytic reduction of NO with C_3H_6, *J. Catal.,* 252, 205–214.

Gobin, O.C. and Schüth, F. (2008). On the suitability of different representations of solid catalysts for combinatoral library design by genetic algorithms, *J. Comb. Chem.*, 10, 835–846.

Goldberg, D. (1989). *Genetic Algorithms in Search, Optimization, and Machine Learning*, Addison-Wesley, Reading, 412 p.

Grubert, G., Kondratenko, E., Kolf, S., Baerns, M., van Geem, P. and Parton, R. (2003). Fundamental insights into the oxidative dehydrogenation of ethane to ethylene over catalytic materials discovered by an evolutionary approach, *Catal. Today*, 81, 337–345.

Grubert, G., Kolf, S., Baerns, M., Vauthey, I., Farrusseng, D., van Veen, A.C., Mirodatos, C., Stobbe, E.R. and Cobden, P.D. (2006). Discovery of new catalytic materials for the water-gas shift reaction by high-throughput experimentation, *Appl. Catal., A: General*, 306, 17–21.

Hendershot, R.J., Rogers, W.B., Snively, C.M., Ogannaike, B.A. and Lauterbach, J. (2004). Development and optimization of NOx storage and reduction catalysts using statistically guided high-throughput experimentation, *Catal. Today*, 98, 375–385.

Holzwarth, A., Denton, P., Zanthoff, H. and Mirodatos, C. (2001). Combinatorial approaches to heterogeneous catalysis: strategies and perspectives for academic research, *Catal. Today*, 67, 309–318.

Huang, K., Zhan, X.L., Chen, F.Q. and Lü, D.W. (2003). Catalyst design for methane oxidative coupling by using artificial neural network and hybrid genetic algorithm, *Chem. Eng. Sci.*, 58, 81–87.

Kennedy, J. and Eberhart R.C. (2001). *Swarm Intelligence,* Morgan Kaufmann, San Francisco, 512 p.

Kirsten, G. and Maier, W.F. (2004). Strategies for the discovery of new catalysts with combinatorial chemistry, *Appl. Surf. Sci.*, 223, 87-101.

Klanner, C., (2004). Evaluation of descriptors for solids, Thesis, Ruhr-University, Bochum, 198 p.

Klanner, C., Farrusseng, D., Baumes, L. Lenliz, M., Mirodatos, C. and Schüth, F. (2004). The development of descriptors for solids: Teaching "catalytic intuition" to a computer, *Angew. Chem. Int. Ed.*, 43, pp. 5347-5349.

Larrañaga, P. and Lozano, J.A. (2002). *Estimation of Distribution Algorithms*, Kluwer, Boston, 382 p.

Li, B., Sun, P., Jin, Q., Wang, J. and Ding, D. (1999). A simulated annealing study of Si, Al distribution in the omega framework, *J. Mol. Catal. A:* Chem., 148, 189–195.

Li, H. Jiao, Y.C. and Wang, Y. (2005). Integrating the simplified interpolation into the genetic algorithm for constrained optimization problems. In: Wang, Y., Cheung, Y. and Liu, H. (eds.), *Computational Intelligence and Security*, Springer, Berlin, pp. 247–254.

Lin, B., Chavali, S., Camarda, K. and Miller, D.C. (2005). Computer-aided molecular design using Tabu search, *Comput. Chem. Eng.*, 29, 337–347.

McLeod, A.S., Johnston, M.E. and Gladden L.F. (1997). Development of a genetic algorithm for molecular scale catalyst design, *J. Catal.*, 167, 279–285.

McLeod, A.S., and Gladden L.F. (2000). Heterogeneous catalyst design using stochastic optimization algorithms, *J. Chem. Inf. Comput. Sci.*, 40, 981–987.

Michalewicz, Z. and Schmidt, M. (2003). Evolutionary Algorithms and Constrained Optimization, In: Sarker, R., Mohammadian, M. and Yao, X. (eds.) *Evolutionary Optimization*, Springer, New York, pp. 57–86.

Michalewicz, Z. and Schonauer, M. (1996). Evolutionary algorithms for constrained parameter optimization problems, *Evol. Comput.*, 4, 1–32.

Mitchell, M. (1996). *An Introduction to Genetic Algorithms,* MIT Press, Cambridge (MA), 224 p.

Nele, M., Vidal, A., Bhering, D.L., Pinto, J.C. and Salim, V.M.M. (1999). Preparation of high loading silica supported nickel catalyst: Simultaneous analysis of the precipitation and aging steps, *Appl. Catal., A: General*, 178, 177–189.

Ohrenberg, A., Törne, C., Schuppert, A. and Knab, B. (2005). Application of data mining and evolutionary optimization in catalyst discovery and high-throughput experimentation — Techniques, strategies, and software, *QSAR Comb. Sci.*, 24, 29–37.

Omata, K., Hashimoto, M., Watanabe, Y., Umegaki, T., Wagatsuma, S., Ishiguro, G. and Yamada, M. (2004). Optimization of Cu oxide catalyst for methanol synthesis under high CO_2 partial pressure using combinatorial tools, *Appl. Catal., A: General*, 262, 207–214.

Otten, R.H.J.M. and Ginneken, L.P.P.P. (1989). *The Annealing Algorithm*, Springer, New York, 224 p.

Paul, J.S., Janssens, R., Denayer, J.F.M., Baron, G.V. and Jacobs, P.A. (2005). Optimization of MoVSb oxide catalyst for partial oxidation of isobutane by combinatorial approaches, *J. Comb. Chem.*, 8, 407–413.

Pereira, R.M., Clerc, F., Farrusseng, D., Waal, J.C. and Maschmeyer, T. (2005). Effect of the genetic algorithm parameters on the optimisation of heterogeneous catalysts, *QSAR Comb. Sci.*, 24, 45–57.

Ramos, R., Menendez, M. and Santamaria, J. (2000). Oxidative dehydrogenation of propane in an inert membrane reactor, *Catal. Today*, 56, 239–245.

Reeves, C.R. and Rowe, J.E. (2003). *Genetic Algorithms: Principles and Perspectives*, Kluwer, Boston, 332 p.

Reid. D.J. (1996). Genetic algorithms in constrained optimization, *Math. Comput. Model.*, 23, 87–111.

Rodemerck, U., Wolf, D., Buyevskaya, O.V., Claus, P., Senkan, S. and Baerns, M. (2001). High-throughput synthesis and screening of catalytic materials: Case study on the search for a low-temperature catalyst for the oxidation of low-concentration propane, *Chem. Eng. J.*, 82, 3–11.

Rodemerck, U., Baerns, M., Holeňa, M. and Wolf, D. (2004). Application of a genetic algorithm and a neural network for the discovery and optimization of new solid catalytic materials, *Appl. Surf. Sci.*, 223, 168–174.

Schaefer, R. (2007). *Foundation of Global Genetic Optimization*, Springer, Berlin, 222 p.

Senkan, S. (2001). Combinatorial heterogeneous catalysis — a new path in an old field, *Angew. Chem. Int. Ed.*, 40, 312–329.

Serra, J.M., Chica, A. and Corma, A. (2003). Development of low temperature light paraffin isomerization catalysts with improved resistance to water and sulphur by combinatorial methods, *Appl. Catal., A: General*, 239, 35–42.

Serra, J.M. and Corma, A. (2004). Two exemplified combinatorial approaches for catalytic liquid-solid and gas-solid processes in oil refining and fine chemicals. In: Hagemeyer, A., Strasser, P. and Volpe, A.F. (eds.), *High-Throughput Screening in Chemical Catalysis*, Wiley-WCH, Weinheim, p. 129–151.

Serra, J.M., Baumes, L.A., Moliner, M., Serna, P. and Corma, A. (2007). Zeolite synthesis modelling with support vector machines: a combinatorial approach, *Comb. Chem. High Throughput Screening*, 10, 13–24.

Snyman, J.A. (2005). *Practical Mathematical Optimization. An Introduction to Basic Optimization Theory and Classical and New Gradient-Based Algorithms*, Springer, New York, 257 p.

Tagliabue, M., Carluccio, L.C., Ghisletti, D. and Perego C. (2003). Multivariate approach to zeolite synthesis, *Catal. Today*, 81, 405–412.

Tompos, A., Margitfalvi, J.L., Tfirst, E., Végvári, L., Jaloull, M.A., Khalfalla, H.A. and Elgarni, M.M. (2005). Development of catalyst libraries for total oxidation of methane: A case study for combined application of "holographic research strategy and artificial neural networks" in catalyst library design, *Appl. Catal., A: General.*, 285, 65–78.

Tompos, A., Végvári, L , Tfirst, E. and Margitfalvi, J.L. (2007). Assessment of predictive ability of artificial neural networks using holographic mapping, *Comb. Chem. High Throughput Screening*, 10, 121–134.

Urschey, J., Kühnle, A., and Maier, W.F. (2003). Combinatorial and conventional development of novel dehydrogenation catalysts, *Appl. Catal., A: General*, 252, 91–106.

Valero, S., Argente, E., Botti, V., Serra, J.M., Serna, P., Moliner, M. and Corma. A. (2009). DoE framework for catalyst development based on soft computing techniques, *Comput. Chem. Eng.*, 33, 225–238.

Végvári, L., Tompos, A., Göbölös, S. and Margitfalvi, J.F. (2003). Holographic research strategy for catalyst library design: Description of a new powerful optimisation method, *Catal. Today*, 81, 517–527.

Vose, M.D. (1999). *The Simple Genetic Algorithm: Foundations and Theory*, MIT Press, Cambridge (MA), 251 p.

Watanabe, Y., Umegaki, T., Hashimoto, M., Omata, K., Yamada, M. (2004). Optimization of Cu oxide catalysts for methanol synthesis by combinatorial tools using 96 well microplates, artificial neural network and genetic algorithm, *Catal. Today*, 89, 455–464.

Wolf, D., Buyevskaya, O.V. and Baerns, M. (2000). An evolutionary approach in the combinatorial selection and optimization of catalytic materials, *Appl. Catal., A: General*, 200, 63–77.

Wong, M.L. and Leung, K.S. (2000). *Data Mining Using Grammar Based Genetic Programming and Applications,* Kluwer, Dordrecht, 213 p.

Yamada, Y. and Kobayashi, T. (2006). Utilization of combinatorial method and high throughput experimentation for development of heterogeneous catalysts, *J. Jpn. Pet. Inst.*, 49, 157–167.

Zabinsky, Z.B. (2003). *Stochastic Adaptive Search for Global Optimization*, Kluwer, Boston, 224 p.

Chapter 4

Generating Problem-Tailored Genetic Algorithms for Catalyst Search

This chapter returns to the topic of genetic algorithms developed specifically for optimisation of catalytic materials and presents a recent approach that preserves the advantages of such algorithms being tailored to catalytic tasks but avoids the disadvantageous necessity to reimplement the algorithm when changes occur in the kind of materials to be optimised. The central idea here is to automatically generate problem-tailored implementations from requirements concerning the materials with a program generator. For the specification of such requirements, a formal description language — called *catalyst description language* (CDL) — has been developed. What can be described by that language will be explained below, in Section 4.2.

4.1. Using a Program Generator — Why and How

As already mentioned in the previous chapter, the difficulty pertaining to specific GAs in catalysis is the restriction of the full usefulness of any specific genetic algorithm implementation to the task for which it was designed. It was this difficulty that inspired the basic idea behind the proposed approach: to postpone the implementation of the algorithm as much as possible, i.e., to only implement the algorithm immediately before it is used to solve a particular optimisation task. Since at that time all requirements concerning the task are already known, this approach enables the GA to be precisely problem-tailored.

ExperimentInformation **GlobalParameter** Pd_catalyst, generation size 45
PopulationSize **GlobalParameter** 45
CurrentExperimentId **GlobalParameter** Pd_cat
ExperimentIdField **GlobalParameter** experiment
GenerationField **GlobalParameter** generation
SimulationFlagField **GlobalParameter** simulation
SequentialNrField **GlobalParameter** sequential_nr
ExternalIdsTable **GlobalParameter** pdcat
ExternalIdsField **GlobalParameter** external_ids
FeedbackTable **GlobalParameter** pdcat
FeedbackField **GlobalParameter** fitness
EvolutionField **GlobalParameter** evolution
OverviewTable **GlobalParameter** overview
Pd_catalyst **ComposedOf** support **InProportion** support_fraction, support_dopants
& InProportion support_dopants_fraction, Pd **InProportion** 0.01
& dopants **InProportion** dopants_fraction **PreparedUsing**
& dopants_preparation
support_fraction **FromInterval** 0.87,1 **WithPrecision** 0.01
support_dopants_fraction **FromInterval** 0,0.2 **WithPrecision** 0.001
dopants_fraction **FromInterval** 0,0.012 **WithPrecision** 0.0001
dopants_preparation **OneOf** 1,2
support_fraction + support_dopants_fraction + 0.01 + dopants_fraction = 1
support **OneOf** silica, alumina
support_dopants **ComposedOf** number_of_support_dopants **FromAmong** B **InProportion**
& B_fraction, Ti **InProportion** Ti_fraction, Ce **InProportion** Ce_fraction, Co
& InProportion Co_fraction, Y **InProportion** Y_fraction, La **InProportion**
& La_fraction, Mo **InProportion** Mo_fraction
number_of_support_dopants **OneOf** 0,1,2
B_fraction **FromInterval** 0.01,0.1 **WithPrecision** 0.001
Ti_fraction **FromInterval** 0.01,0.1 **WithPrecision** 0.001
Ce_fraction **FromInterval** 0.01,0.1 **WithPrecision** 0.001
Co_fraction **FromInterval** 0.01,0.1 **WithPrecision** 0.001
Y_fraction **FromInterval** 0.01,0.1 **WithPrecision** 0.001
La_fraction **FromInterval** 0.01,0.1 **WithPrecision** 0.001
Mo_fraction **FromInterval** 0.01,0.1 **WithPrecision** 0.001
B_fraction + Ti_fraction + Ce_fraction + Co_fraction + Y_fraction +
& La_fraction + Mo_fraction = support_dopants_fraction
dopants **ComposedOf** number_of_dopants **FromAmong** Mg **InProportion** Mg_fraction, V
& InProportion V_fraction, Mo_dopant **InProportion** Mo_dopant_fraction, Mn
& InProportion Mn_fraction, Co_dopant **InProportion** Co_dopant_fraction, Rh
& InProportion Rh_fraction, Ni **InProportion** Ni_fraction, Ag **InProportion**
& Ag_fraction, Zn **InProportion** Zn_fraction, Sn **InProportion** Sn_fraction
number_of_dopants **OneOf** 0,1,2
Mg_fraction **FromInterval** 0,0.01 **WithPrecision** 0.0001
V_fraction **FromInterval** 0,0.01 **WithPrecision** 0.0001
Mo_dopant_fraction **FromInterval** 0,0.01 **WithPrecision** 0.0001
Mn_fraction **FromInterval** 0,0.01 **WithPrecision** 0.0001
Co_dopant_fraction **FromInterval** 0,0.01 **WithPrecision** 0.0001
Rh_fraction **FromInterval** 0,0.01 **WithPrecision** 0.0001
Ni_fraction **FromInterval** 0,0.01 **WithPrecision** 0.0001
Ag_fraction **FromInterval** 0,0.01 **WithPrecision** 0.0001

Figure 4.1. Simple example of a complete CDL description. Description keywords are in boldface.

```
Zn_fraction FromInterval 0,0.01 WithPrecision 0.0001
Sn_fraction FromInterval 0,0.01 WithPrecision 0.0001
Mg_fraction + V_fraction + Mn_fraction + Rh_fraction + Ni_fraction +
& Ag_fraction + Zn_fraction + Sn_fraction + Co_dopant_fraction +
& Mo_dopant_fraction = dopants_fraction
silica IsPrimitive
alumina IsPrimitive
Pd IsPrimitive
B IsPrimitive
Ti IsPrimitive
Ce IsPrimitive
Co IsPrimitive
Y IsPrimitive
La IsPrimitive
Mo IsPrimitive
Mg IsPrimitive
V IsPrimitive
Mn IsPrimitive
Rh IsPrimitive
Ni IsPrimitive
Ag IsPrimitive
Zn IsPrimitive
Sn IsPrimitive
Mo_dopant IsPrimitive
Co_dopant IsPrimitive
support_fraction SavedIn pdcat
support_dopants_fraction SavedIn pdcat
dopants_fraction SavedIn pdcat
dopants_preparation SavedIn pdcat
support SavedIn pdcat
B_fraction SavedIn pdcat
Ti_fraction SavedIn pdcat
Ce_fraction SavedIn pdcat
Co_fraction SavedIn pdcat
Y_fraction SavedIn pdcat
La_fraction SavedIn pdcat
Mo_fraction SavedIn pdcat
Mg_fraction SavedIn pdcat
V_fraction SavedIn pdcat
Mo_dopant_fraction SavedIn pdcat
Mn_fraction SavedIn pdcat
Co_dopant_fraction SavedIn pdcat
Rh_fraction SavedIn pdcat
Ni_fraction SavedIn pdcat
Ag_fraction SavedIn pdcat
Zn_fraction SavedIn pdcat
Sn_fraction SavedIn pdcat
number_of_support_dopants SavedIn pdcat
number_of_dopants SavedIn pdcat
```

Figure 4.1. (*Continued*)

However, traditional implementation undertaken by human programmers is not feasible in such a situation since it is error-prone, expensive and too slow. It is proposed therefore that problem-tailored GA implementations be *generated automatically*. To this end, a *program generator* is required, i.e., a software system that converts given requirements into executable programs. In contrast to a human programmer, for a program generator the requirements have to be expressed in a rigorously formal way. For this purpose, we use a Catalyst Description Language, developed at LIKAT (Holeňa, 2004; Holeňa *et al.*, 2008, cf. Figure 4.1).

An overall schema for the proposed approach is set out in Figure 4.2. The program generator parses text files with CDL descriptions as input, and produces GA implementations as output.

For this approach, it is immaterial what the program generator looks like. It can be programmed in various languages, and it can be a standalone program or can combine calls to generic GA software with parts implementing the functionality that the generic software does not cover.

If the values of the objective function have to be obtained through experimental testing, the GA implementation runs only once and then exits. However, the approach also previews the possibility of obtaining those values from a simulation program instead (cf. Chapter 7). In such a case, the GA implementation alternates with that program for as many generations as desired.

4.2. Description Language for Optimisation Tasks in Catalysis

The proposed Catalyst Description Language allows the specification of a broad variety of user requirements to the catalytic materials to be sought by the genetic algorithm, as well as to the algorithm itself. Most important among those requirements are the following:

a) Which substances should form the pool from which the various *components of the catalytic material* are selected.
b) In which *hierarchy* the components from the pool should be organized (Figure 4.3). CDL allows the specified hierarchy (called

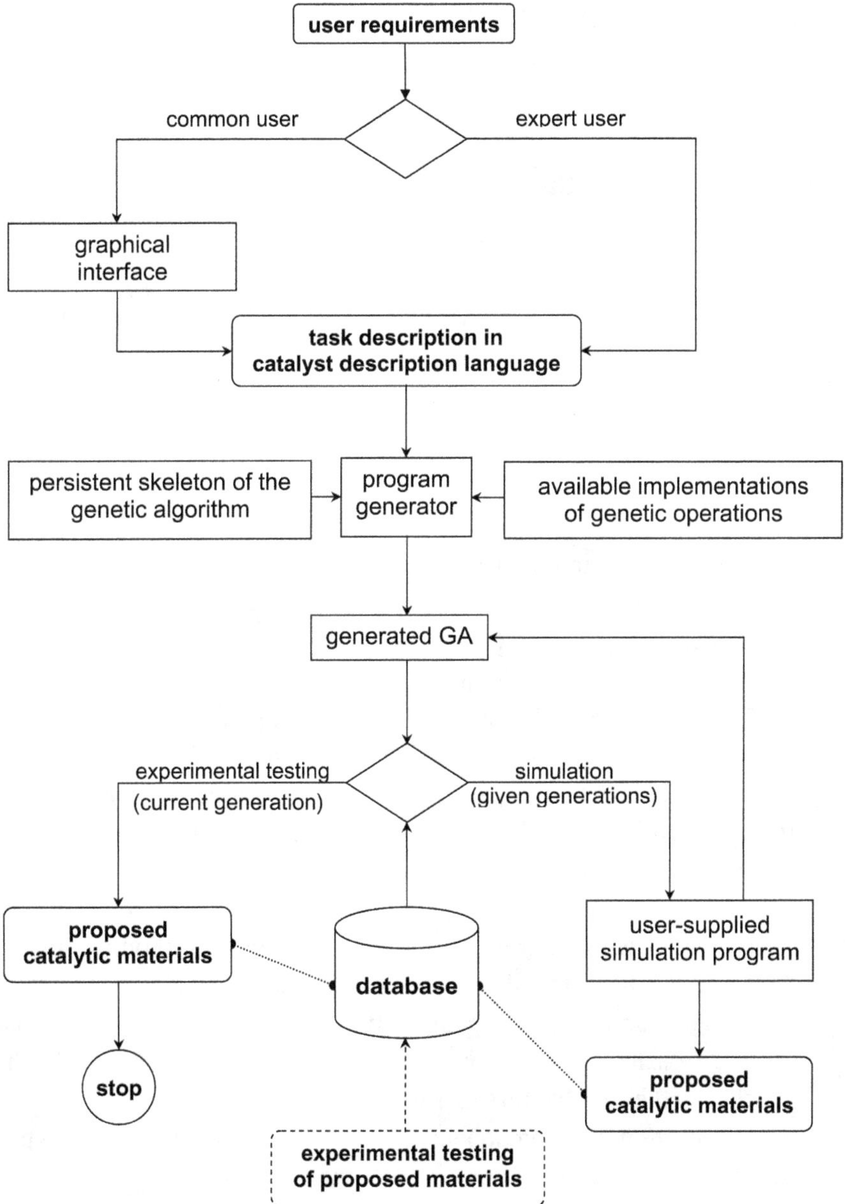

Figure 4.2. Schema of the functionality of a program generator that generates problem-tailored genetic algorithms according to CDL descriptions.

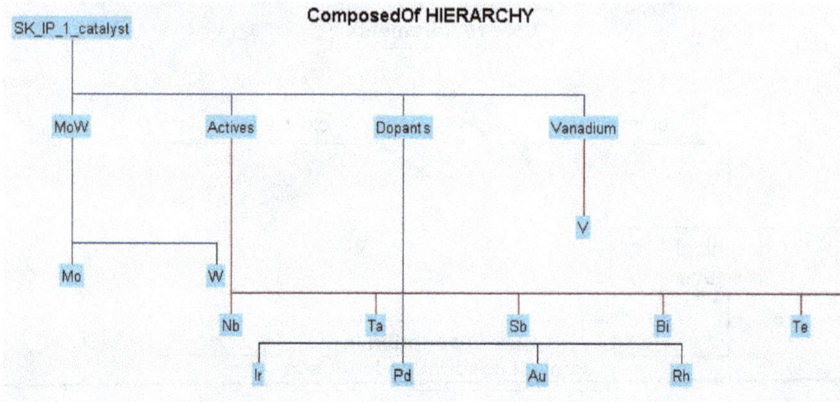

Figure 4.3. Example hierarchy of component types, as depicted in the graphical user interface for creating CDL descriptions.

"ComposedOf hierarchy") to be arbitrarily complex. At the highest level, it contains general types of components, such as active components, support or dopants. Each of them may have its own subtypes, those again subsubtypes, etc.

c) The number of components that may be *simultaneously present* in an individual material, as well as the number of simultaneously present components of a particular type from within the component-types hierarchy (e.g., the material should contain five components altogether, three of which are active components).

d) The choice among several possibilities for the *number of simultaneously present components* (e.g., the material should contain 4–6 components altogether, of which 2–4 should belong to active components and 0–2 to dopants). By default, all possibilities occur with the same probability, but this can be changed through specifying their probability distribution.

e) The *proportion* of a component or a type in the catalyst, or in a type at a higher level of the hierarchy (e.g., the proportion of the support in the catalyst, or the proportion of a particular active component among all active components).

f) A *lower and an upper bound for the proportion* of a component (e.g., the proportion of the support in the catalyst is 75–80%, or the proportion of a particular active component among active components is 20–100%).
g) In addition to the quantities introduced so far — i.e., proportions, numbers of simultaneously present elements, and lower or upper bounds — also *any additional quantities*, to describe various chemical or physical properties of the catalysts, their preparation procedure and reaction conditions, and also the behaviour of the genetic algorithm itself. For these additional quantities, lower and upper bounds or probability distributions of the values of the quantity may also be specified.
h) *Linear equality or inequality constraints* for any quantities (e.g., proportion of Mg among all active components + proportion of Mn among all active components = 50%, lower bound for the proportion of active components > five times the upper bound for the proportion of dopants). Each such constraint may contain an arbitrary number of quantities, and each quantity may occur in an arbitrary number of constraints.
i) The *choice among several possibilities* for the preparation of the catalyst, of any component, and of any type from within the component-types hierarchy. In addition, any step of the preparation (e.g., precipitation, calcination), and any of its features can again be chosen from among several possibilities. In this way, the component types hierarchy is complemented by an arbitrarily complex hierarchy of choices. Moreover, that hierarchy can also be applied to quantities (cf. d).
j) *Population size* of the current generation.
k) *Descriptive information* about the current experiment provided by the user.
l) The particular way in which the *genetic operations* selection, crossover and mutation should be performed.

<table>
<tr><td>CatalystDescription ::=</td><td>ComposedOfClause [linefeed CatalystDescriptionClause]*</td></tr>
<tr><td>CatalystDescriptionClause ::=</td><td>{ ComposedOfClause
| Identifier FromInterval IntervalBound, IntervalBound WithPrecision Precision
| GlobalParameterIdentifier GlobalParameter Value
| Identifier IsPrimitive
| Identifier OneOf SelectableIdentifier [,SelectableIdentifier]* [DistributedAs Distribution]
| KnownIdentifier SavedIn Table [InField Field]
| [real number*] Quantity [{ + | – } [real number*] Quantity]* Comparison [real number*] Quantity }</td></tr>
<tr><td>ComposedOfClause ::=</td><td>Identifier ComposedOf [Count FromAmong]
KnownComponent [InProportion Quantity]
[PreparedUsing KnownIdentifier]
[,KnownComponent [InProportion Quantity]
[PreparedUsing KnownIdentifier]]*</td></tr>
<tr><td>Identifier ::=</td><td>letter [{ letter | digit }]*, if this string is not a Keyword</td></tr>
<tr><td>Keyword ::=</td><td>{ Always | ComposedOf | DistributedAs
| FromAmong | FromField | FromInterval
| GlobalParameter| InField
| InProportion | IsPrimitive
| OneOf| PreparedUsing | SavedIn
| WithParameters| WithPrecision}</td></tr>
<tr><td>KnownIdentifier ::=</td><td>{ KnownComponent
| EvolvableIdentifier
| Quantity
| Count
| Identifier1 in Identifier1 IsPrimitive }</td></tr>
<tr><td>KnownComponent ::=</td><td>{ Identifier1 in Identifier1 ComposedOf [Count FromAmong]
KnownComponent [InProportion Quantity]
[PreparedUsing KnownIdentifier]
[,KnownComponent [InProportion Quantity]
[PreparedUsing KnownIdentifier]]*
| Identifier2 in Identifier2 IsPrimitive
| Identifier3 in Identifier3 one of
KnownComponent [,KnownComponent]*
[DistributedAs Distribution] }</td></tr>
<tr><td>Quantity ::=</td><td>{ real number
| Identifier1 in Identifier1 FromInterval IntervalBound, IntervalBound WithPrecision Precision
[WithParameters *ParameterString*]]
| Identifier2 in Identifier2 OneOf Quantity, Quantity]* [DistributedAs Distribution]</td></tr>
<tr><td>Distribution ::=</td><td>real number, real number [,real number]*</td></tr>
<tr><td>IntervalBound ::=</td><td>[real number *] *Quantity*</td></tr>
</table>

Figure 4.4. Syntax of the proposed Catalyst Description Language (CDL).

<table>
<tr><td>GlobalParameterIdentifier :: =</td><td>{ ExperimentInformation
| PopulationSize
| CurrentExperimentId
| ExperimentIdField
| GenerationField
| SequentialNrField
| SimulationFlagField
| ExternalIdsTable
| FeedbackTable
| FeedbackField
| EvolutionField
| EvolutionMethod
| EvolutionParameters
| OverviewTable }</td></tr>
<tr><td>SelectableIdentifier::=</td><td>{ real number
| Identifier1 in Identifier1 IsPrimitive
| Identifier2 OneOf SelectableIdentifier [,SelectableIdentifier]*
[DistributedAs Distribution]}</td></tr>
<tr><td>EvolvableIdentifier ::=</td><td>{ Quantity in Quantity FromInterval IntervalBound,
IntervalBound WithPrecision Precision
[WithParameters ParameterString]]
| Identifier1 in KnownComponent ComposedOf
Count FromAmong KnownComponent [InProportion
Quantity] PreparedUsing Identifier1
[,KnownComponent [InProportion Quantity]
[PreparedUsing Identifier]]*
| Identifier2 in Identifier2 OneOf SelectableIdentifier
[,SelectableIdentifier]* [DistributedAs Distribution]}</td></tr>
<tr><td>Count ::=</td><td>{ nonnegative integer
| Identifier1 in Identifier1 OneOf Count [,Count]*
[DistributedAs Distribution] }</td></tr>
<tr><td>Value ::=</td><td>{ real number
| string
| Identifier1 in Identifier1 IsPrimitive
| Table
| Field
| EvolutionMethod }</td></tr>
<tr><td>Precision ::=</td><td>a power of 0.1</td></tr>
<tr><td>DistributionMethod ::=</td><td>name of an accessible Matlab function</td></tr>
<tr><td>ParameterString ::=</td><td>string</td></tr>
<tr><td>Table ::=</td><td>name of a valid database table</td></tr>
<tr><td>Field ::=</td><td>name of a valid field of a valid database table</td></tr>
<tr><td>Comparison ::=</td><td>{ = | ~= | < | <= | > | >= }</td></tr>
</table>

Figure 4.4. (*Continued*)

m) Which particular parts of the algorithm output should be *stored in the database*, and in which *tables and fields* should they be stored. It is necessary to store in the database at least those parts of the algorithm output that are vital for correct functioning of the algorithm in subsequent generations. Consequently, the tables and fields must be specified at least for these parts of the output, which include in particular:
   - All output obtained using *any random choices*, such as the choice of the specified number of components from the pool, the choice among several possibilities, or the choice of the value of a quantity from the interval between a lower and an upper bound, since such output cannot be reconstructed if it has not been stored.
   - The information needed to *uniquely identify a catalytic material in the database*, which consists of a unique database identifier for the experiment considered, the number of the current generation, and the number of the catalyst within that generation.

n) *Degreee of precision* with which any part of the algorithm output should be stored in the database. In this way, the generated algorithm, during its search for new catalytic materials, will then avoid finding those materials that within the given degree of precision, already exists in the database.

o) Which table and field of the database contain the *actual values of the objective function*.

The formal CDL syntax is summarised in Figure 4.4. A detailed explanation of that syntax can be found in the research report by Holeňa (2007).

Once the objective function has been chosen (e.g., yield, degree of conversion), then the optimisation task is fully determined as soon as we specify the set on which the optimum of that function should be searched. To specify this set in turn requires specifying:

(i) The *meaning of the individual coordinates* of points in the set, i.e., the meaning of coordinates in the input space of the objective function in which this set lies. Recall from the previous section that in catalysis, the coordinates convey the qualitative and quantitative composition of the catalyst and its preparation and reaction conditions. The correspondence between them and the requirements expressible in CDL is established in Table 4.1.

(ii) *Delimitation of the set* on which the optimum should be searched within the space where it lies. The above overview of requirements expressible in CDL indicates several possible ways of delimiting that set, in particular in terms of:

- Linear *equality or inequality constraints* for quantities.
- The *number of components simultaneously present* in the catalyst, as well as numbers of simultaneously present components of particular types, possibly together with the choice of their values among several possibilities, and sometimes also together with their probability distribution on the set of possible values.
- The *lower and upper bounds for proportions.*

Table 4.1. Correspondence between the meaning conveyed by coordinates of points in the input space of the objective function and the requirements expressible in CDL.

Meaning conveyed by coordinates	Requirement expressible in CDL
Qualitative composition of the catalyst	Pool of catalyst components, hierarchy of component types
Quantitative composition of the catalyst	Proportion of a component or a type in the catalytic material, or a in a component type on a higher level of the component-types hierarchy
Catalyst preparation	Choice among several possibilities for the preparation of the catalyst, of a component or a type, or choice among several possibilities for a preparation step or feature
Reaction conditions of the catalyzed reaction	Additional quantities, together with a choice among several possibilities for their values, optionally with their lower and upper bounds

4.3. Tackling Constrained Mixed Optimisation

Mixed constrained optimisation problems were in general discussed in the previous chapter. To address them in the generated genetic algorithms, we make use of two specific features of such problems in the area of catalytic materials:

(i) It is sufficient to consider only *linear constraints*. Even if the set of feasible solutions is not constrained linearly in reality, the finite measurement precision of the continuous variables involved always allows one to constrain it piecewise linearly. The relevant linear piece can then be indicated using an additional discrete variable. Consequently, the set of feasible values of the continuous variables that form a solution together with a particular combination of values of the discrete variables is a *polyhedron*, although each such polyhedron can be empty, and each has its specific dimension, ranging from 1 (closed interval) to the number of continuous variables.

(ii) If a solution polyhedron is described with an inequality
$\boldsymbol{P} = \{\boldsymbol{x} : \boldsymbol{A}\boldsymbol{x} \leq \boldsymbol{b}\}$,
then its feasibility — i.e., the property $\boldsymbol{P} \neq \boldsymbol{\emptyset}$ — is invariant with respect to any permutation of columns of $\boldsymbol{A}$, as well as with respect to any permutation of rows of $(\boldsymbol{A}\ \boldsymbol{b})$. Moreover, since identity is also a permutation, since each permutation has a unique inverse permutation, and since the composition of permutations is again a permutation, the relation $\approx$ between solution polyhedra, defined for $\boldsymbol{P}$ and $\boldsymbol{P}' = \{\boldsymbol{x} : \boldsymbol{A}'\boldsymbol{x} \leq \boldsymbol{b}'\}$ by means of

$\boldsymbol{P} \approx \boldsymbol{P}'$ iff $(\boldsymbol{A}'\ \boldsymbol{b}')$ can be obtained from $(\boldsymbol{A}\ \boldsymbol{b})$ through
some permutation of columns of A, followed by (1)
some permutation of rows of the result and of $\boldsymbol{b}$

is an *equivalence relation* on the set of solution polyhedra. Consequently, this relation partitions that set into disjoint *equivalence classes*, the number of which can be much lower than the number of polyhedra. In the optimisation tasks faced during the testing of the approach, the number of solution polyhedra ranged between thousands and hundreds of thousands, but the number of their equivalence classes ranged only between several and several dozens. As an example, the simple catalyst optimisation task described in Figure 4.1 leads to 6494 solution polyhedra, which form nine equivalence classes with the equivalence relation (1). A more realistic example is documented in Figure 4.5, using the graphical interface with status information for the algorithms generated by prototype implementation of the program generator, which is outlined in Section 4.4. In this case there are 583232160 solution polyhedra, which form 480 equivalence classes. Whereas separately checkimg the non- emptiness of each polyhedron could prohibitively increase computing time for the generated GA, forming the equivalence classes is fast, and then only one representative from each class needs to be checked.

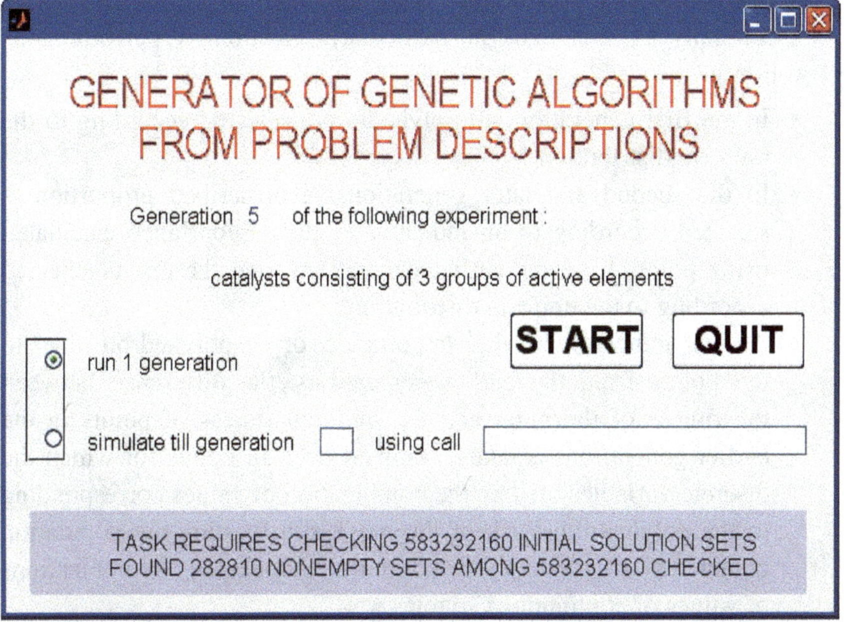

Figure 4.5. Simple graphical interface with status information for the generated GA.

Combining the above approach with the fact that the constraints determine which combinations of values of discrete variables to consider, one may suggest the following algorithm for dealing with constrained mixed optimisation in the generated GA:

1. A separate continuous optimisation problem is formulated for each combination of values of discrete variables that can be for some combination of continuous variables feasible with respect to the specified constraints.
2. The set of all solution polyhedra corresponding to the continuous optimization problems formulated in step 1 is partitioned according to the above equivalence.
3. One polyhedron from each partition class is checked for non-emptiness, taking into account the discernibility (measurement precision) of variables considered.
4. On the set of non-empty polyhedra, discrete optimisation is performed, using operations selection and mutation developed specifically to this end. In particular, selection is performed as follows:
   - In the first generation, all polyhedra are selected according to the uniform distribution.
   - In the second and later generations, a prescribed proportion is selected according to an indicator of their importance calculated using points from the earlier generations, and the rest is selected according to the uniform distribution.
   - As an indicator of the importance of a polyhedron due to the points from the earlier generations, the difference between the fitness of the point and the minimal fitness of points in the earlier generations is taken, summed over all points for which the discrete variables assume the combination of values corresponding to the polyhedron. Each of the polyhedra forming the population obtained in this way corresponds to a subpopulation of combinations of values of continuous variables.

   Through the discrete optimisation, values of discrete variables are fixed, such as the numbers of components of individual types in the catalytic material or the methods of preparation of its individual parts.
5. In each of the polyhedra found through the discrete optimisation, a continuous optimisation is performed. Through the continuous optimisation, values of continuous variables, i.e., of proportions of individual components and component types in the material, are

fixed. The combinations of values of continuous variables found in this way, combined with the combinations of values of discrete variables corresponding to the respective polyhedra, together form the final population of solutions to the mixed-optimisation problem.

4.4. A Prototype Implementation

As was already mentioned in Section 4.1, it is immaterial how a program generator underlying the above approach is implemented. It is not the purpose of this chapter to discuss different possibilities for its implementation or to advertise any particular implementation. For illustration only, a prototype implementation in a system called GENACAT will be briefly outlined, which has been implemented at LIKAT (Figure 4.6). The system has been developed in Matlab, and also the genetic operations, as well as the skeleton, i.e., the persistent part of the generated GA, are implemented in Matlab. In particular, they make use of the Matlab Genetic Algorithm and Direct Search Toolbox (2004), as well as the Multi-Parametric Toolbox from ETH Zurich (Kvasnica *et al.*, 2005). On the other hand, the variable part of the GA is generated as a binary code, which is input to the GA skeleton.

To *switch between a single run of the genetic algorithm and simulation*, and set the name of the simulation program and its parameters, the GA generator provides a simple graphical inerface (Figure 4.5). In addition, that interface continuously shows the number of catalysts proposed so far, the current phase of the entire optimisation process and other status information.

Much more complex is the graphical interface that allows users to enter the information needed to create a CDL description. Its main window is shown in Figure 4.6. In addition, the interface provides the possibility of visualising the hierarchy of component types and the hierarchy of choices (Figure 4.3). The decision whether to use this interface or to write the CDL description manually is made in the introductory window of GENACAT. Moreover, that introductory window also allows the use of an existing description and the introduction of necessary changes within it, as well as storing

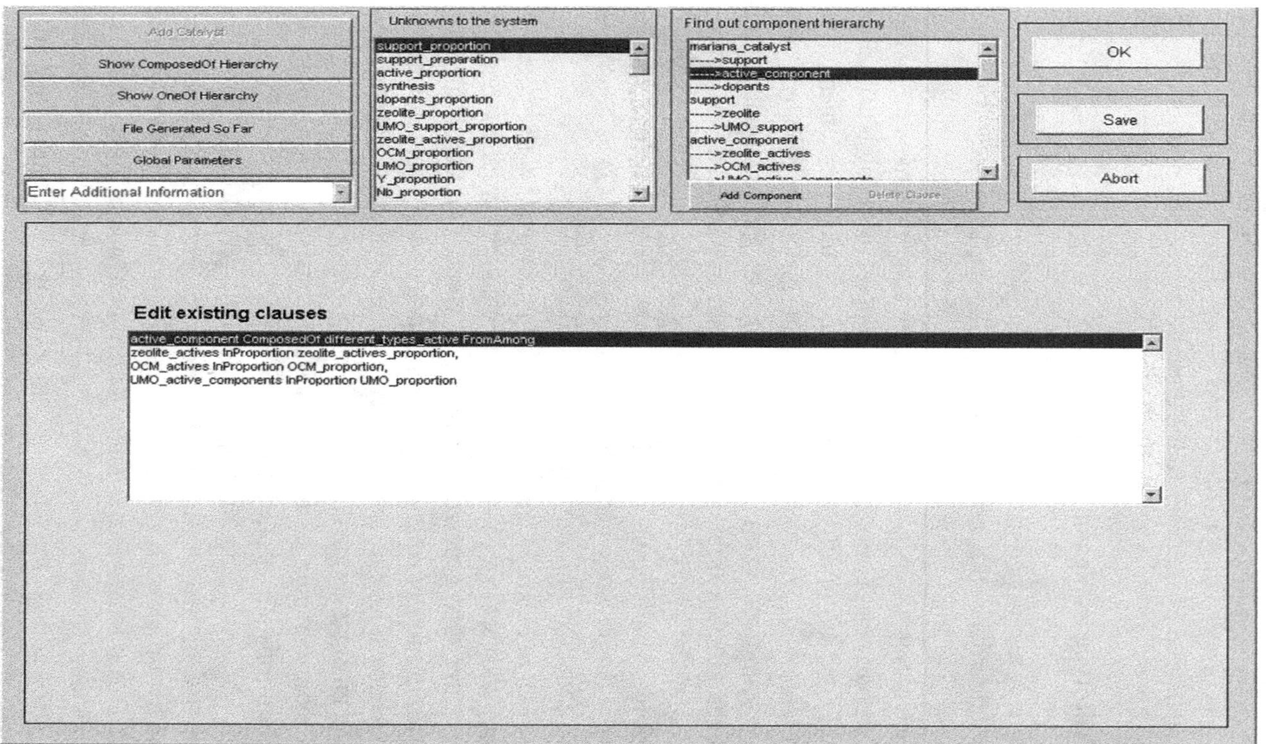

Figure 4.6. Main window of the graphical öf GENACAT interface allowing users to enter the information needed to create a CDL description.

fragmentary descriptions for later reopening and completing. These features have a twofold importance. First, much time can be saved in preparing the description if the intended experiment is similar to an earlier one. Second, the user does not become bored with the necessity of dealing with information that lies outside his or her area of competence and that can actually be provided by a colleague. This is intended in the first place for information concerning the database and simulations, which is typically provided by a database administrator or a data analyst, but it also allows, for example, to provide information about catalyst composition and about preparation methods by two independent experimenters.

Bibliography

Genetic Algorithm and Direct Search Toolbox (2004). The MathWorks, Inc., Natick, 268 p.

Holeňa, M. (2004). Present trends in the application of genetic algorithms to heterogeneous catalysis, In: Hagemeyer, A., Strasser, P. and Volpe, A.F. (eds.), *High-Throughput Screening in Chemical Catalysis*, Wiley-WCH, Weinheim, pp. 53–172.

Holeňa, M. (2007). *Description Language for Catalyst Search with Evolutionary Methods*, Leibniz Institute for Catalysis, Berlin, 14 p.

Holeňa, M., Čukić, T., Rodemerck, U. and Linke, D. (2008). Optimization of catalysts using specific, description-based genetic algorithms, *J. Chem. Inf. Model.*, 48, 274–282.

Kvasnica, M., Grieder, P., Baotic, M. and Christophersen, F.J. (2005). *Multi-Parametric Toolbox (MPT)*, ETH, Zurich, 105 p.

Chapter 5

Analysis and Mining of Data Collected in Catalytic Experiments

5.1. Similarity and Difference Between Data Analysis and Mining

The term *data analysis* is traditionally used for statistically assessing the extent to which data support particular relationships between various objects, properties and phenomena that they describe. Untill the early 1970s, researchers hypothesised such relationships, and computers were employed only for their validation. However, the advent of databases made it possible to collect increasingly huge amounts of experimental data which could not be dealt with in this traditional way. Instead, methods for automatically searching such relationships in data started to be developed within a new branch of statistics, called *exploratory statistics*. On the other hand, the development of computers also stimulated the emergence of methods that attempted to mimic the way in which humans get knowledge from empirical data. This group of methods, generally referred to as *machine learning*, belongs to the area of artificial intelligence and relies on results from both mathematics (logic, information theory, probability theory, approximation theory) and other areas (biology, neurology, psychology). Although methods of exploratory statistics and machine-learning methods rely on different paradigms, they deal with the same data and employ the same information technologies — especially databases, the World-Wide Web and various object-oriented technologies (cf. Figure 5.1). Together, they form a new interdisciplinary area called *data mining*. In the context of heterogeneous catalysis, the distinction between traditional data analysis and data mining should be taken into account when choosing between the two conceptually different approaches to the design of libraries of solid catalysts — the optimisation approach and the discovery approach (Klanner *et al.*, 2003).

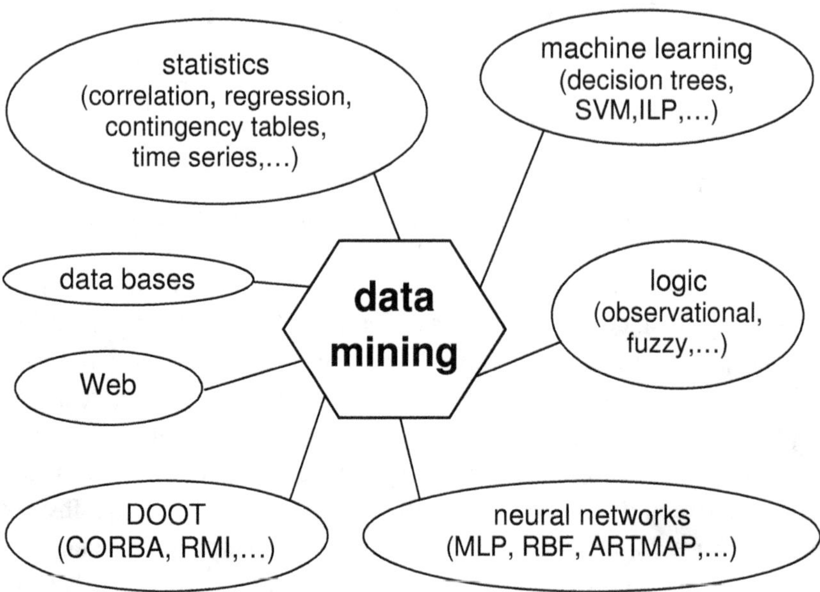

Figure 5.1. Main data mining approaches and supporting technologies (abbreviations: ILP — inductive logic programming, MLPs — multilayer perceptrons, RBF — radial basis functions networks, ARTMAP — adaptive resonance theory mapping networks, DOOT — distributed object-oriented technologies, CORBA — common object request broker, RMI — remote method invocation).

5.2. Survey of Existing Methods

In this overview section, both traditional statistical methods and the more recent machine-learning methods are briefly surveyed. Excluded here is only the main tool for data analysis and data mining of catalytic materials, i.e., the application of artificial neural networks, to which two of the remaining chapters will be devoted.

5.2.1. *Statistical Methods*

The simplest use of statistical methods is to provide summary parameters characterising important statistical properties of input variables and of various measures of catalyst performance (such as yield or degree of conversion), or relationships between them. Such summary parameters are usually called *descriptive statistics*; their common representatives are *mean*, *median*, *variance*, *standard deviation*, *covariance* and *correlation*.

The last two descriptive statistics mentioned, covariance and correlation, allow to summarise the relationship between a performance measure and a particular input variable. The situation gets substantially more complicated if one is interested in the relationship between a performance measure and a whole set of input variables. Indeed, in that case, not only parameters corresponding to individual variables, but also parameters corresponding to various levels of interactions between them are needed. In such situations, the parameters are usually combined with an assumption about the form of the dependency of the performance measure on the input variables. Most commonly assumed is linear dependency, polynomial dependency, dependency based on a generalisation of the scalar product in an approach called support vector machines, or dependency derived from some theoretical model. Once an assumption about its form is made, the parameters already fully determine that dependency. In statistics, this approach is called *regression* or *response surface modelling*, and the parameters determining the dependency are called *regressors*. For example, the regression of yield $\boldsymbol{y}$ on three component fractions $\boldsymbol{x}_1$, $\boldsymbol{x}_2$, $\boldsymbol{x}_3$ uses in the case of a *linear regression* four parameters $\boldsymbol{a}_0$, $\boldsymbol{a}_1$, $\boldsymbol{a}_2$, $\boldsymbol{a}_3$:

$$\boldsymbol{y} = \boldsymbol{a}_0+\boldsymbol{a}_1\boldsymbol{x}_1+\boldsymbol{a}_2\boldsymbol{x}_2+\boldsymbol{a}_3\boldsymbol{x}_3,$$

whereas in the case of a *quadratic regression*, it uses ten parameters $\boldsymbol{a}_0$, $\boldsymbol{a}_1$, $\boldsymbol{a}_2$, $\boldsymbol{a}_3$, $\boldsymbol{a}_{1,1}$, $\boldsymbol{a}_{1,2}$,..., $\boldsymbol{a}_{3,3}$:

$$\boldsymbol{y} = \boldsymbol{a}_0+\boldsymbol{a}_1\boldsymbol{x}_1+\boldsymbol{a}_2\boldsymbol{x}_2+\boldsymbol{a}_3\boldsymbol{x}_3+ +\boldsymbol{a}_{1,1}\boldsymbol{x}_1^2+\boldsymbol{a}_{2,2}\boldsymbol{x}_2^2+\boldsymbol{a}_{3,3}\boldsymbol{x}_3^2+\boldsymbol{a}_{1,2}\boldsymbol{x}_1\boldsymbol{x}_2$$
$$+\boldsymbol{a}_{1,3}\boldsymbol{x}_1\boldsymbol{x}_3+\boldsymbol{a}_{2,3}\boldsymbol{x}_2\boldsymbol{x}_3.$$

Common to all regression models is a difference between the character of the variables on the left side and on the right side of the equation describing the model:

- On the right side, there are values of variables that are chosen by the experimenter, most importantly descriptors characterising the composition of catalytic materials and their properties and preparation, but also variables characterising reaction conditions. These variables are called *input variables* or *independent variables*, and their choice forms the core of experiment design.
- On the left side, there are values of variables that describe various aspects of catalyst performance, e.g., yield of particular reaction products, degree of conversion of a particular feed component, or selectivity with respect to a particular product. These variables are results of the experiment and are closely related to its objective, which can be either the search for catalysts with optimal performance, or the investigation of the dependence of particular aspects of catalyst performance on particular input variables. They are referred to as *output variables* or *dependent variables*.

In the most typical case where the independent variables are describing the composition of the catalytic material, the obtained regression model is frequently termed *quantitative structure-activity relationship* (QSAR), cf. Klanner *et al.* (2004), Farrusseng *et al.* (2005), Ohrenberg *et al.* (2005), Farrusseng *et al.* (2007), Serra *et al.* (2007).

Descriptive statistics are also not sufficient in the case of other properties of catalytic materials that depend on a whole set of input variables. Examples of such properties are similarity between different materials, dependence on unobservable variables (also called latent variables), or classification of materials according to their catalytic behavior in particular reactions. To characterise such properties, methods of *multivariate analysis* are needed. Multivariate approaches most relevant to the analysis of catalytic data are:

(i) *Principal component analysis*, which reduces data dimensionality through concentrating on those linear combinations of input variables that are most responsible for the variability of the data set (unfortunately, however, particular linear combinations usually do not convey any real meaning).

(ii) *Factor analysis*, which explains input variables as combinations of a smaller number of unobservable factors, and indicates which part of the values of dependent variables is determined by other variables and which part is due to noise.

(iii) Analysis of the relationship between factors influencing input variables and those influencing performance measures by means of an approach called *partial least squares*.

(iv) *Cluster analysis*, grouping catalytic materials into clusters or a hierarchy of clusters according to similarity among the values of the input variables or according to a similar catalytic performance, which can be measured using various *similarity measures*.

(v) *Classification* of new materials, according to values of their input variables, with respect to their usability as catalytic materials in particular reactions. There are various approaches to statistical classification, such as linear and quadratic discriminant analysis, nearest neighbour approach, Bayesian approach, classification trees, or support vector machines. Their common feature is that the discrimination between classes relies solely on data with known correct classification, for example, on already tested materials.

Frequently, the primary purpose of data analysis is to check the compatibility of the available data against certain assumptions about the probability distribution governing those data. To this end, methods of *statistical hypotheses testing* are needed. Most commonly, hypotheses such as the following are tested:

- The probability distribution belongs to a certain family of distributions, for example normal distributions or exponential distributions.
- Certain parameters characterising the distribution, such as mean or variance, have particular values, or their values are within particular ranges.
- Probability distributions governing two datasets are identical, or some of their parameters are identical.

An important example of a hypothesis testing of the last-mentioned kind is the analysis of the influence of varying the values of individual input variables (e.g., catalyst components and their proportions) on the performance of a catalytic material by means of an approach called *analysis of variance*. The approach assumes that each output variable follows some basic statistical model, in which the expectation of the output variable is viewed as the sum of the effects of individual input variables (called main effects), possibly superimposed by their interactions of various complexity (Scheffé, 1999; Sahai and Ageel, 2000). The amount of available data for each combination of values of input variables determines how complex the basic model will be. The principle of the approach consists in testing the hypothesis that a particular main effect or interaction can be left out of that model without significantly changing the value of the output variable. If the tested hypothesis is valid, then both models will give the same error. Therefore, the ratio of both errors is computed in the analysis-of-variance method, and if that ratio differs significantly from the value 1, the tested hypothesis is rejected. Provided that the individual errors are normally distributed, also the distribution of the error ratio is known (this is called the Fisher-Snedecor distribution). Using this distribution, the probability can be computed that the error ratio is as high as the value corresponding to the measured data, or even higher. That probability is called *achieved significance* of the test. The lower it is, the more unlikely it is that the measured data could occur if the simplified model is valid. Consequently, the more significant is then the effect/interaction that was left out of the model.

For detailed information about statistical methods, the reader is again referred to comprehensive monographs in statistics (Hastie *et al.*, 2001; Berthold and Hand, 2002; Schölkopf and Smola, 2002; Kantardzic, 2003), or in chemometrics (Meier and Zund, 1993; Caria, 2000). Separately, we would like to draw readers' attention to monographs on an important nonlinear regression method related to artificial neural networks, which form the topic for the next chapter — namely, to kernel-based support vector regression (Schölkopf and Smola, 2002; Steinwart and Christmann, 2008). The support vector approach is actually a more general approach, based on the statistical theory of learning (Vapnik, 1995; Hastie *et al.*, 2001), which also pertains to cluster analysis and especially to classification. The first applications of this approach to catalysis have recently been reported (Baumes *et al.*, 2006; Serra *et al.* 2007). Examples of other recent applications of statistical methods to heterogeneous catalysis are Richardson *et al.* (2003), Urschey *et al.* (2003), Wilkin *et al.* (2003), Amat *et al.* (2004), Cawse *et al.* (2004), Hendershot *et al.* (2004), Klanner (2004), Klanner *et al.* (2004), Cantrell *et al.* (2005), Corma *et al.* (2005), Farrusseng *et al.* (2005), Hendershot *et al.* (2005), Scheidtmann *et al.* (2005) , Serra *et al.* (2007) and Sieg *et al.* (2007).

5.2.2. *Extraction of Logical Rules from Data*

Probably the oldest data-mining approach outside the area of exploratory statistics is the extraction of logical rules from data by means of *observational logic*, which is a Boolean predicate logic with generalised quantifiers (Hájek and Havránek, 1978). Most frequently, *association rules* are extracted from data, i.e., rules of observational logic that correspond to generalised quantifiers expressing a high conditional probability of the consequence of a rule if its condition is valid (Zhang and Zhang, 2002). For example, the following rule could be extracted from data collected in experiments with the synthesis of hydrocyanic acid (cf. Section 5.3):

(support = Si_3N_4 | support = SiC) & weight proportion of Pt in the loading > 65% $\rightarrow_{0.9}$ Y_{HCN} > 60%

In this rule (as well as in all important kinds of logical rules extracted from data), three parts with different semantics can be distinguished:

- *Condition* (also called *antecedent*), i.e., "support = Si_3N_4 | support = SiC) & weight proportion of Pt in the loading > 65%", which is a formalised notation of the statement that the support of the catalytic material considered is either Si_3N_4 or SiC, and the the weight proportion of Pt in the metal loading exceeds 65%.
- *Consequence* (also called *succedent*), i.e., "Y_{HCN} > 60%", which is a formalised notation of the statement that the HCN yield exceeds 60%.
- *Quantifier*, i.e., "$\rightarrow_{0.9}$", which is one of many generalised quantifiers in observational logic. The particular quantifier $\rightarrow_p$ employed in the above rule is called *founded implication* and expresses the fact that among those objects in the available data that fulfil the condition of the rule, at least the proportion $\boldsymbol{p}$ also fulfils its consequence (in the above rule, $\boldsymbol{p}$ = 90%).

Consequently, the rule as a whole expresses the fact that in the available data, at least 90% of catalytic materials with support Si_3N_4 or SiC and the proportion of Pt in the metal loading over 65% led to an HCN yield exceeding 60%.

When extracting rules of the observational logic or association rules, there is no obstacle if the antecedents of two or more rules overlap. On the contrary, a substantial proportion of the data often fulfils the antecedents of several rules simultaneously. However, rules with overlapping antecedents are undesirable if they are intended to be used for classification, and especially if they are intended to serve as a basis for decision making. In such situations, methods producing sets of Boolean implications with non-overlapping antecedents are used; the rules from such sets are called *decision rules*. The best known representatives of methods for the extraction of decision rules are the methods denoted *AQ* (Michalski 1980; Michalski and Kaufmann, 2001) and *CN2* (Clark and Boswell, 1991), and especially a large group of methods called *decision trees* (Breiman *et al.*, 1984; Quinlan, 1992; Hastie *et al.*, 2001). The term used for the latter reflects the fact that rulesets obtained by means of such methods have a hierarchical structure, so they can easily be visualised as tree-like graphs. It is this easy visualizability of the sets of rules obtained that accounts for the considerable popularity of this class of methods. Moreover, decision trees are quite robust against outliers since the borderlines between areas corresponding to antecedents of different rules are piecewise-constant and do not depend on distant data. The principle of the method consists in splitting the value set of each input variable in such a way that the sum of the empirical variances of the output variable computed for data in both partitions is minimised. In this way, the method forms a hierarchy of partitions of the value sets for the input variables. Depending on the number of such splits that are consecutively performed, trees of different sizes are obtained. The most appropriate tree size is usually chosen using a method called cross-validation. Since this method is also used for finding the most appropriate size of artificial neural networks, it will be explained in connection with the latter in Chapter 8.

Inductive logic programming (ILP) consists principally in inducing a relation defined by means of logical rules, from positive and negative examples of that relation in data, while other similarly defined relations may be used as background knowledge in the induced definition (De Raedt, 1992; Muggleton, 1992). Nowadays, a variety of ILP systems exists, for the induction of a single concept or multiple concepts, in batch mode or in incremental mode, and for interactive or non-interactive induction.

Observe that all machine learning methods mentioned so far extract knowledge from data in the *symbolic form* of rules. Nevertheless, this is not a necessary feature of all such methods. Indeed, in the important group of methods relying on artificial neural networks, knowledge is extracted in the *numeric form* of *quantitative relationships between variables*. Since exactly such relationships are of our primary concern in catalysis, it is not surprising that methods based on neural networks are encountered most frequently in the search of catalytic materials — even more frequently than methods of statistical origin. Among rule-extraction methods, only decision trees and the extraction of association rules have occasionally been employed in this area (Klanner, 2004; Klanner *et al.*, 2004; Clerc *et al.*, 2005; Corma *et al.*, 2005; Farusseng *et al.*, 2005, Cukic *et al.*, 2005; Moehmel *et al.*, 2008; Rothenberg, 2008). An introduction to artificial neural networks from the viewpoint of data mining in catalysis will be given in the next chapter. In that chapter, also the little-known fact will be recalled and illustrated that they underlie not only methods extracting from data quantitative relationships, but also some rule-extraction methods.

Both data analysis and data mining are performed in a stepwise manner; the result of a particular step might then be used in later steps. This is particularly apparent if the objective of an initial stage of data analysis or data mining is to find out which input variables play the most important role in the relationships investigated, or which of them share their roles in that relationship. Such an initial stage is called *feature selection*, and leads ultimately to *dimensionality reduction* of the space of input variables, subsequently allowing a more thorough analysis in that less-dimensional space (Klanner, 2004; Klanner *et al.*, 2004; Corma *et al.*, 2005).

5.3. Case Study with the Synthesis of HCN

One statistical and one machine-learning method are illustrated in a case study using data from the investigation of catalytic materials for the *high-temperature synthesis of hydrocyanic acid.* This investigation and its results were recently described in Moehmel *et al.* (2008). The investigation was performed through high-throughput experiments in a circular 48-channel reactor. In most of these experiments, the composition of the materials was designed using the first genetic algorithm developed specifically for the optimisation of solid catalysts (Buyevskaya *et al.*, 2000; Wolf *et al.*, 2000; Buyevskaya *et al.*, 2001; Rodemerck *et al.*, 2001). More precisely, the algorithm ran for seven generations with a population of 92 catalytic materials, and in addition 52 other catalysts were investigated, with composition designed directly by the experimenter, without using the genetic algorithm. Consequently, data from a total of 696 catalytic materials were collected. The methods employed in this case study were the analysis of variance and decision trees.

The composition and preparation of the catalytic materials studied and the conditions to which they had been exposed have been described in detail in Moehmel *et al.* (2008). Here, only those facts are recalled that are important for understanding which variables are considered in this case study.

(i) All the materials tested contained a *support*, which was sequentially coated with one to six layers of active metal additives. As support, in every case one of the following *15 materials* was used: pure α-Al_2O_3 (alsint), as well as the compounds AlN, Mo_2C, TiB_2, TiN, NB_2O_3, BN, ZrO_2, Sm_2O_3, SrO, CaO, MgO, TiO_2, SiC, and Si_3N_4, bound in an aluminium matrix.

(ii) The active layers by which the supports were covered had been selected from the pool of 11 compounds: $Y(NO_3)_3$, $La(NO_3)_3$, $ZrO(NO_3)_2$, H_2MoO_4, $ReCl_3$, $IrCl_4$, $NiCl_2$, H_2PtCl_6, $Zn(NO_3)_2$, $AgNO_3$, and $HAuCl_4$, providing in turn the 11 metal additives Y, La, Zr, Mo, Re, Ir, Ni, Pt, Zn, Ag and Au. It is important to realise that the proportions of these compounds in the active part of the catalyst are not completely independent since they sum up to 100%.

(iii) The catalytic performance, in particular the *degrees of conversion of CH_4 and NH_3*, and the *HCN yield*, was measured only after the final composition of the inlet feed gas was reached through stepwise addition of CH_4 to the initial ammonia/argon mixture. That final composition amounted to 10.7 vol. % NH_3, 9.3 vol. % CH_4 and 80 vol. % Ar. The catalytic performance was then measured for *temperatures in the range 1173K–1373K*. Consequently, it is not possible to simply use all the performance measurement results as values of output variables — either the measurement temperature has to be added to the input variables, or only results corresponding to a single temperature can be used. The latter possibility was preferred and only data collected at the highest temperature 1373K were analysed.

In the context of the analysis of variance, these facts imply that the expectation of HCN yield, $\boldsymbol{E}Y_{\mathrm{HCN}}$, equals the sum of main effects, $\alpha_{\mathrm{support}} + \alpha_{\mathrm{Y}} + \ldots + \alpha_{\mathrm{Au}}$, to which interactions of two variables (e.g., $\alpha_{\mathrm{support,Y}} + \alpha_{\mathrm{support,La}} + \alpha_{\mathrm{Y,La}} \ldots + \alpha_{\mathrm{Ag,Au}}$) or even of more variables (e.g., $\alpha_{\mathrm{support,Y,La}} + \ldots$) can be added. This is the basis model from which are derived hypotheses about the possibility of leaving out a particular main effect or interactions, according to the preceding section. For example, in case of the basic model $\boldsymbol{E}Y_{\mathrm{HCN}} = \alpha_{\mathrm{support}} + \alpha_{\mathrm{Y}} + \ldots + \alpha_{\mathrm{Au}}$, the hypothesis that the effect of support could be left out of the model means this model could actually be simplified to $\boldsymbol{E}Y_{\mathrm{HCN}} = \alpha_{\mathrm{Y}} + \ldots + \alpha_{\mathrm{Au}}$. In the case of a basic model considering only main effects, the following data are needed:

- Data about at least one material with each of the 15 possible supports.
- For each metal additive, data about at least one material in which that additive is present and at least one in which it is absent.

Similarly, in the case of a basic model with two-variables interactions, the following data are needed for each combination of support and metal additive:

- Data about at least one material, in which that combination is present.
- Data about at least one material, in which the support is present but the additive is absent.

The results of the analysis of variance are presented in Table 5.1. The basic model included only main effects because the available data are not sufficient for a model with interactions — some combinations of support and metal additives are not available. Consequently, only the influence of a single variable (support, the presence of a particular metal additive) on HCN yield has been tested. Most significant was the influence of the support and Pt as active metal component. Both parameters were very significant (the probability defining the achieved significance level was very low). This means the support and the presence or absence of Pt will strongly influence HCN formation. Significant influence on HCN formation (achieved significance level less than 0.4%) was also observed for the presence or absence of the metal additives Ir, Mo, Zn, Au and Ni. On the opposite end, the achieved significance level of 98.9% indicates that the presence or absence of Zr has no influence on HCN yield. A relatively low influence on HCN formation is also indicated for the presence or absence of the metal additives Ag and Re.

Figure 5.2 shows the decision tree obtained from the available data. The metal loadings labelling the branches of the tree are related to the

Table 5.1. Results of testing the individual main effects in the analysis of variance.

Variable	Achieved significance (%)
Support	$< 10^{-14}$
Pt	$< 10^{-14}$
Ag	32.6
Au	0.1
Ir	$2{\cdot}10^{-6}$
La	4.6
Mo	$2{\cdot}10^{-4}$
Ni	0.4
Re	16.0
Y	3.0
Zn	0.2
Zr	98.9

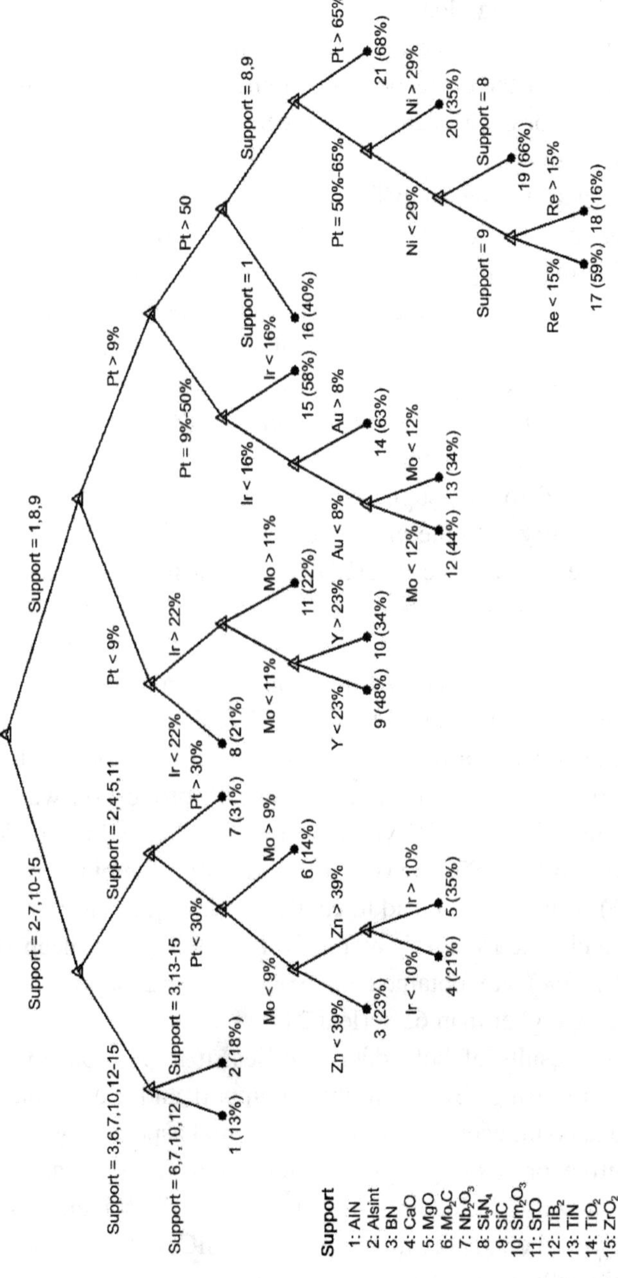

Figure 5.2. Regression tree obtained for the regression of HCN on the support and on the relative amounts of 11 metal additives (in parentheses behind the leaf numbers, the mean value of HCN yield for the catalysts belonging to that leaf; values without brackets give the relative metal loading related to the total metal loading of 2.2 weight %).

total amount of the metal loading (which was 2.2 weight %). As an example, a value of > 65% means that the catalyst is loaded with more than 1.44 weight % of the corresponding element. The leaves of this tree are numbered, and for each of them the mean value of the HCN yield is given (in parentheses).

In a similar way to the analysis of variance, the decision tree also reveals that the most important input variable in order to achieve high yields of HCN is the catalyst support. Using the supports 2, 3, 4, 5, 6, 7, 10, 11, 12, 13, 14, 15 (the meaning of the individual nummbers is explained in Table 5.2), the mean value of HCN yield is at most 35% (leaves 1–7). With the supports 3 (BN), 6 (Mo_2C), 7 (Nb_2O_3), 10 (Sm_2O_3), 12 (TiB_2), 13 (TiN), 14 (TiO_2) and 15 (ZrO_2), only catalysts with low mean value of HCN yield are obtained (leaves 1, 2). The mean values of HCN yield for the supports 2 (Alsint), 4 (CaO), 5 (MgO), 11 (SrO) increase slightly with the proportion of Pt (leaf 31), and of Ir (leaf 35). However, the mean value of HCN formation for these supports is comparatively low. The supports 1 (AlN), 8 (Si_3N_4) and 9 (SiC) mostly deliver catalysts of high HCN yield (leaves 8–21). With these supports, active phases or species for HCN formation are formed at their surface after heating the catalytic material to 1373K and stepwise addition of methane. The mean value of the yield of HCN for these supports depends mainly on the proportion of Pt amount. If the Pt proportion was lower than 9%, a mean value of HCN yield below 48% was achieved (leaves 8–11). For a proportion of Pt between 9 and 50%, combinations of Pt with Au (> 8%) or Ir (> 16%) lead to catalysts with mean HCN yields of 63% (leaf 14) and 58% (leaf 15), respectively. The highest mean values of HCN yield (68%) are obtained on Si_3N_4 and SiC supports with a relative Pt loading higher than 65% (leaf 21).

Based on the results of both data analyses, it is possible to design future experiments using fewer variables (support and metal additives) and taking into account economic or environmental aspects, e.g. costs for the metal additive or toxicity. For the development of an active and selective catalyst with a Pt loading lower than 50%, the screening should be carried out with the support materials Si_3N_4, SiC and AlN containing the metal additives Pt, Ir and Au.

Table 5.2. Support materials and their BET surface area.

No.	support	BET surface (m^2/g)
1	AlN	n.d.
2	Alsint	0.7
3	BN	0.2
4	CaO	0.3
5	MgO	2.7
6	Mo_2C	n.d.
7	Nb_2O_3	0.16
8	Si_3N_4	0.7
9	SiC	1
10	Sm_2O_3	0.2
11	SrO	0.1
12	TiB_2	0.7
13	TiN	n.d
14	TiO_2	0.15
15	ZrO_2	0.3

Bibliography

Amat, A.M., Arques, A., Bossmann, S.H., Braun, A.M., Göb, S., Miranda, M.A. and Oliveros, E. (2004). Oxidative degradation of 2,4-xylidine by photosensitization with 2,4,6-triphenylpyrylium: Homogeneous and heterogeneous catalysis, *Chemosphere*, 57, 1123–1130.

Baumes, L.A., Serra, J.M., Serna, P. and Corma, A. (2006). Support vector machines for predictive modelling in heterogeneous catalysis: A comprehensive introduction and overfitting estimation based on two real applications, *J. Comb. Chem.*, 8, 583–596.

Berthold, M. and Hand, D.J. (2002). *Intelligent Data Analysis: An Introduction*, 2nd Ed., Springer, Berlin, 460 p.

Breiman, L., Friedman, J.H., Olshen, R.A. and Stone, C.J. (1984). *Classification and Regression Trees*, Wadsworth, Belmont, 358 p.

Buyevskaya, O.V., Wolf, D. and Baerns, M. (2000). Ethylene and propene by oxidative dehydrogenation of ethane and propane: Performance of rare-earth oxide-based catalysts and development of redox-type catalytic materials by combinatorial methods, *Catal. Today*, 62, 91–99.

Buyevskaya, O.V., Bruckner, A., Kondratenko, E.V., Wolf, D. and Baerns, M. (2001). Fundamental and combinatorial approaches in the search for and optimization of catalytic materials for the oxidative dehydrogenation of propane to propene, *Catal. Today*, 67, 369–378.

Cantrell, D.G., Gillie, L.J., Lee, A.F. and Wilson, K. (2005). Structure-reactivity correlations in MgAl hydrotalcite catalysts for biodiesel synthesis, *Appl. Catal., A: General*, 287, 183–190.

Caria, M. (2000). *Measurement Analysis. An Introduction to the Statistical Analysis of Laboratory Data in Physics, Chemistry and the Life Sciences*, Imperial College Press, London, 229 p.

Cawse, J.N., Baerns, M. and Holeňa, M. (2004). Efficient discovery of nonlinear dependencies in a combinatorial data set, *J. Chem. Inf. Comput. Sci.*, 44, 143–146.

Clark, P. and Boswell, R. (1991). Rule induction with CN2: Some recent improvements, *Machine Learning — European Working Session on Learning*, Springer, Berlin, pp. 151–163.

Čukić, T., Kraehnert, R. Holeňa, M., Herein, D., Linke, D. and Dingerdissen, U. (2007). The influence of preparation variables on the performance of Pd/Al_2O_3 catalyst in the hydrogenation of 1,3-butadiene: Building a basis for reproducible catalyst synthesis, *Appl. Catal., A: General,* 323, 25–37.

De Raedt, L. (1992). *Interactive Theory Revision: An Inductive Logic Programming Approach*, Academic Press, London, 256 p.

Farrusseng, D., Klanner, C., Baumes, L., Lengliz, M., Mirodatos, C. and Schüth, F. (2005). Design of discovery libraries for solids based on QSAR models, *QSAR Comb. Sci.*, 24, 78–93.

Farrusseng, D., Clerc, F., Mirodatos, C., Azam, N., Gilardoni, F., Thybaut, J.W., Balasubramaniam, P. and Marin, G.B. (2007). Development of an integrated informatics toolbox: HT kinetic and virtual screening, *Comb. Chem. High Throughput Screening*, 10, 85–97.

Hájek, P. and Havránek, T. (1978). *Mechanizing Hypothesis Formation*, Springer, Berlin, 396 p.

Hastie, T., Tibshirani, R. and Friedman, J. (2001). *The Elements of Statistical Learning*, Springer, Berlin, 552 p.

Hendershot, R.J., Rogers, W.B., Snively, C.M., Ogannaike, B.A. and Lauterbach, J. (2004). Development and optimization of NO_x storage and reduction catalysts using statistically guided high-throughput experimentation, *Catal. Today*, 98, 375–385.

Hendershot, R.J., Vijay, R., Feist, B.J., Snively, C.M. and Lauterbach, J. (2005). Multivariate and univariate analysis of infrared imaging data for high-throughput studies of NH_3 decomposition and NOx storage and reduction catalysts, *Meas. Sci. Technol.*, 16, 302–308.

Kantardzic, M. (2003). *Data Mining. Concepts, Models, Methods, and Algorithms*, Wiley, Chichester, 360 p.

Klanner, C., Farrusseng, D., Baumes, L. Mirodatos, C. and Schüth, F. (2003). How to design diverse libraries of solid catalysts? *QSAR Comb. Sci.*, 22, 729–736.

Klanner, C. (2004). Evaluation of descriptors for solids, Thesis, Ruhr-University, Bochum, 198 p.

Klanner, C., Farrusseng, D., Baumes, L., Lenliz, M., Mirodatos, C. and Schüth, F. (2004). The development of descriptors for solids: teaching "catalytic intuition" to a computer, *Angew. Chem. Int. Ed.*, 43, 5347–5349.

Meier, P.C. and Zund, R.E. (1997). *Statistical Methods in Analytical Chemistry*, Wiley, New York, 456 p.

Michalski, R.S. (1980). Knowledge acquisition through conceptual clustering: A theoretical framework and algorithm for partitioning data into conjunctive concepts, *Int. J. Policy Anal. Inf. Syst.*, 4, 219–243.

Michalski, R.S. and Kaufman, K.A. (2001). Learning patterns in noisy data. In: Porioulos, G., Karkaletsis, G. and Spyropoulos, C.O. (eds.), *Machine Learning and Its Applications*, Springer, Berlin, pp. 22–38.

Moehmel, S., Steinfeldt, N., Engelschalt, S., Holeňa, M., Kolf, S., Baerns, M., Dingerdissen, U., Wolf, D., Weber, R. and Bewersdorf, M. (2008). New catalytic materials for the high-temperature synthesis of hydrocyanic acid from methane and ammonia by high-throughput approach, *Appl. Catal., A: General*, 334, 73–83.

Muggleton, S. (1992). *Inductive Logic Programming*, Academic Press, London, 565 p.

Ohrenberg, A., Törne,C., Schuppert, A. and Knab, B. (2005). Application of data mining and evolutionary optimization in catalyst discovery and high-throughput experimentation — Techniques, strategies, and software, *QSAR Comb. Sci.*, 24, 29–37.

Quinlan, J. (1992). *C4.5: Programs for Machine Learning*, Morgan Kaufmann, San Francisco, 302 p.

Richardson, J.T., Remue, D. and Hung, J.K. (2003). Properties of ceramic foam catalyst supports: Mass and heat transfer, *Applied Catalysis A: General*, 250, 319–329.

Rodemerck, U., D. Wolf, D., Buyevskaya, O.V., Claus, P., Senkan, S. and Baerns, M. (2001). High-throughput synthesis and screening of catalytic materials: Case study on the search for a low-temperature catalyst for the oxidation of low-concentration propane, *Chem. Eng. J.*, 82, 3–11.

Rothenberg, G. (2008). Data mining in catalysis: Separating knowledge from garbage, *Catal. Today*, 137, 2–10.

Sahai, H. and Ageel, M.I. (2000). *Analysis of Variance: Fixed, Random and Mixed Models*, Birkhäuser, Boston, 784 p.

Scheidtmann, J., Klaer, D., Saalfrank, J.W., Schmidt, T. and Maier, W.F. (2005). Quantitative composition activity relationships (QCAR) of Co-Ni-Mn-mixed oxide and M1-M2-mixed oxide catalysts, *QSAR Comb. Sci.*, 24, 203–210.

Scheffé, H. (1999). *The Analysis of Variance*, Wiley, New York, 477 p.

Schölkopf, B. and Smola, A.J. (2002). *Learning with Kernels*, MIT Press, Cambridge (MA), 644 p.

Serra, J.M., Baumes, L.A., Moliner, M., Serna, P. and Corma, A. (2007). Zeolite synthesis modelling with support vector machines: A combinatorial approach, *Comb. Chem. High Throughput Screening*, 10, 13–24.

Sieg, S.C., Suh, C., Schmidt, T., Stukowski, M., Rajan, K. and Maier, W.F. (2007). Principal component analysis of catalytic functions in the composition space of heterogeneous catalysts, *QSAR Comb. Sci.*, 26, 528–535.

Steinwart, I. and Christmann, A. (2008). Support Vector Machines, Springer, Berlin, 601 p.

Urschey, J., Kühnle, A., and Maier, W.F. (2003). Combinatorial and conventional development of novel dehydrogenation catalysts, *Appl. Catal., A: General*, 252, 91–106.

Vapnik, V. (1998). *Statistical Learning Theory,* Wiley, Chichester, 736 p.

Wilkin, O.M., Maitlis, P.M., Haynes, A. and Turner, M.L. (2003). Mid-IR spectroscopy for rapid on-line analysis in heterogeneous catalyst testing. *Catal. Today*, 81, 309–317.

Wolf, D., Buyevskaya, O.V. and Baerns, M. (2000). An evolutionary approach in the combinatorial selection and optimization of catalytic materials, *Appl. Catal., A: General*, 200, 63–77.

Zhang, C. and Zhang, S. (2002). *Association Rule Mining: Models and Algorithms*, Springer, Berlin, 238 p.

Chapter 6

Artificial Neural Networks in the Development of Catalytic Materials

6.1. What are Artificial Neural Networks?

In the preceding chapter, it was mentioned that statistics offers regression models for modelling a simultaneous dependency of a variable on several other variables. However, there is also an important class of machine learning methods that suit the same purpose — feed-forward neural networks, which are the most frequently encountered representatives of artificial neural networks (ANNs). From a statistical viewpoint, feed-forward neural networks actually compute complicated nonlinear regression models. However, they originated outside statistics, and are typically used without any statistical setting. This is particularly true for their applications in catalysis (Kito *et al.*, 1994; Hattori and Kito, 1995; Sasaki *et al.*, 1995; Hou *et al.*, 1997; Huang *et al.*, 2001; Liu *et al.*, 2001; Corma *et al.*, 2002; Holeňa and Baerns, 2003; Tompos *et al.*, 2003; Umegaki *et al.*, 2003; Baumes *et al.*, 2004; Omata *et al.*, 2004; Kito *et al.*, 2004; Klanner *et al.*, 2004; Watanabe *et al.*, 2004; Diaconescu and Dumitriu, 2005; Farusseng *et al.*, 2005; Moliner *et al.*, 2005; Omata *et al.*, 2005; Tatliera er al., 2005; Zahedi *et al.*, 2005; Tompos *et al.*, 2006; Serra *et al.*, 2007; Günai and Yildirim, 2008; Valero *et al.*, 2009). They are more frequent than applications of all traditional statistical regression models. For this reason, artificial neural networks will now be presented separately from, and in more detail than, the statistical methods presented in the previous chapter.

The area of artificial neural networks is a highly interdisciplinary area, combining knowledge from computer science, mathematics and biology. This area, which plays a key role in the analysis of experimental data in catalysis, emerged mainly as a consequence of the following influences:

- Mathematical modelling of neurons and neural systems.
- Connectionism, which views large networks of simple elements changing their state through mutual interactions as a crucial condition for the existence of any intelligence.
- Discrepancy between the sequential and algorithmic method of information processing in traditional computers, and the parallel and associative way in which information is processed in biological systems, as well as between the digital representation of information used in the former, and the analogue representation typical for the latter.
- Universal parallel computers and general methods of parallel computation.

Merging all these sources led to the development of artificial neural networks as *distributed computing systems attempting to implement* a greater or smaller part of the *functionality characterising biological neural networks.*

An important property of artificial neural networks is that in contrast to the traditional statistical methods briefly surveyed in the preceding chapter, their applicability does not rely on any specific, hard-to-fulfil conditions. It is worth mentioning that this property also characterises evolutionary algorithms, which were dealt with in Chapters 3 and 4. Mathematical methods that require only weak applicabilty conditions are generally termed *soft computing*. Besides evolutionary algorithms and ANNs, their main representatives are methods based on *fuzzy logic* (Gottwald, 1993; Hájek, 1998; Novák *et al.*, 1999) and on *rough sets* (Pawlak, 1991).

6.1.1. *Network Architecture*

Undoubtedly the most basic concepts pertaining to artificial neural networks are those of a *neuron*, whose biologically inspired meaning is "some elementary signal processing unit", and of a *connection* between neurons enabling the transmission of signals between them. The connections allow one to define, for each neuron $\boldsymbol{v}$, the set $\boldsymbol{i}(\boldsymbol{v})$, *input set* of $\boldsymbol{v}$, as the set of all neurons $\boldsymbol{u}$ from which a connection $(\boldsymbol{u},\boldsymbol{v})$ exists, and the set $\boldsymbol{o}(\boldsymbol{v})$, *output set* of $\boldsymbol{v}$, as the set of all neurons $\boldsymbol{u}$ to which a connection $(\boldsymbol{v},\boldsymbol{u})$ exists.

In actual fact, the connections between neurons cannot be completely arbitrary. Besides the requirement that any connected neurons must be different (no loops), they are usually required to fulfil the following condition, called *condition of non-redundancy*:

For each neuron $\boldsymbol{v}$, at least one of the sets $\boldsymbol{i}(\boldsymbol{v})$ and $\boldsymbol{o}(\boldsymbol{v})$ is nonempty.

In addition to signal transmissions between different neurons, signal transmissions between neurons and the environment can also take place. The set of neurons, connections between them, and connections between some of the neurons and the environment is usually called the *architecture* of the artificial neural network.

The connections between neurons and the environment cannot be fully arbitrary either. Usually, such connections are restricted by the following two complementary requirements:

- Each neuron that sends signals to other neurons must first have received some signal from one or more other neurons or from the environment.
- Each neuron that has received signals from other neurons must subsequently send some signal to one or more other neurons or to the environment.

The architecture of a neural network induces two partitions in the set of its neurons. One of these corresponds to different possibilities of receiving signals:

$\boldsymbol{I}_1$ = {neurons that receive signals from other neurons, but not from the environment},

$\boldsymbol{I}_2$ = {neurons that receive signals from the environment, but not from other neurons},

$\boldsymbol{I}_3$ = {neurons that receive signals both from other neurons, and from the environment}.

The other partition corresponds to different possibilities of sending signals:

$\boldsymbol{O}_1$ = {neurons that send signals to other neurons, but not to the environment},

$\boldsymbol{O}_2$ = {neurons that send signals to the environment, but not to other neurons},

$\boldsymbol{O}_3$ = {neurons that send signals both to other neurons, and to the environment}.

The parts $\boldsymbol{I}_2$ and $\boldsymbol{O}_2$ of these partitions can be simply expressed using the input set $\boldsymbol{i}(v)$ and output set $\boldsymbol{o}(v)$ of a neuron v:

$\boldsymbol{I}_2 = \{v : \boldsymbol{i}(v) \text{ is empty}\}$, $\boldsymbol{O}_2 = \{v : \boldsymbol{o}(v) \text{ is empty}\}$.

Moreover, in terms of these two partitions, the condition of non-redundancy can be reformulated simply as

$\boldsymbol{I}_2$ and $\boldsymbol{O}_2$ are disjoint.

In addition to the above-mentioned requirements, another requirement usually restricts the architecture of an artificial neural network, namely

$\boldsymbol{I}_3$ and $\boldsymbol{O}_3$ are empty.

Whereas the former requirements are inspired by real, biological neural networks, the latter one has been introduced for purely technical reasons — to make a formal description of the artificial neural network easier. On the other hand, that condition does not actually impose any real restriction on the ANN, in the sense that any architecture in which it does not hold can be extended to an architecture in which it does hold, while between any two neurons from the original, non-extended architecture, a connection exists in the extended architecture if and only if it already exists in the original one.

In the case of a neural network architecture in which $\boldsymbol{I}_3$ and $\boldsymbol{O}_3$ are empty, the set $\boldsymbol{I}_2$ is usually denoted only $\boldsymbol{I}$, and its elements are referred to as *input neurons* of the network. Similarly, the set $\boldsymbol{O}_2$ is usually denoted only $\boldsymbol{O}$, and its elements are referred to as *output neurons* of the network. The remaining neurons — those from the intersection $\boldsymbol{H} = \boldsymbol{I}_1 \cap \boldsymbol{O}_1$ of $\boldsymbol{I}_1$ and $\boldsymbol{O}_1$ — are called *hidden neurons* (Figure 6.1).

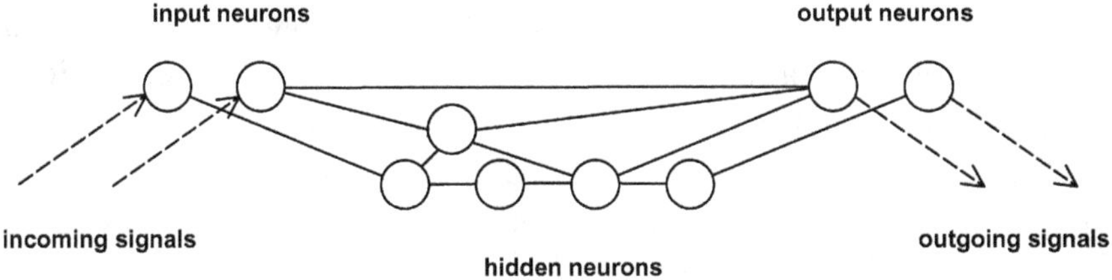

Figure 6.1. A simple generic artificial neural network architecture.

Although the partition of the set of neurons into input, hidden and output neurons still allows a wide variety of architectures, the architecture of nearly all types of artificial neural networks more frequently encountered in practical applications is essentially the same, namely a *layered architecture.* This type of architecture is characterised by the following properties:

- The set of neurons $\boldsymbol{V}$ is partitioned into $\boldsymbol{L}+1$ layers $\boldsymbol{V}_0,\boldsymbol{V}_1,\ldots,\boldsymbol{V}_L$, in such a way that $\boldsymbol{V}_0 = \boldsymbol{I}$, $\boldsymbol{V}_L = \boldsymbol{O}$. Hence if $\boldsymbol{H}$ is nonempty, then $\boldsymbol{L} > 1$ and the layers $\boldsymbol{V}_1,\ldots,\boldsymbol{V}_{L\text{-}1}$ partition $\boldsymbol{H}$, they are called hidden layers.
- If two neurons u and v are connected, then u belongs to some layer $\boldsymbol{V}_k$, $\boldsymbol{k} = 0,\ldots,\boldsymbol{L}\text{-}1$ whereas v belongs to $\boldsymbol{V}_{k+1}$. If in addition all neurons of any layer $\boldsymbol{V}_k$, $\boldsymbol{k} = 0,\ldots,\boldsymbol{L}\text{-}1$ are connected to all neurons of the following layer $\boldsymbol{V}_{k+1}$, then the layered architecture is called *fully connected.*

In spite of being actually partitioned into $\boldsymbol{L}+1$ layers, a neural network with such an architecture is conventionally called an ***L**-layer network* (due to the fact that signals undergo transformations only in the layers of hidden and output neurons, not in the input layer). In particular, a *one-layer network* is a layered neural network without hidden neurons, whereas a *two-layer network* is a neural network in which only connections from input to hidden neurons and from hidden to output neurons are possible.

6.1.2. *Important Kinds of Neural Networks*

The most prominent example of one-layer networks are *perceptrons* — the earliest representative of artificial neural networks in the sense outlined above (proposed in Rosenblatt, 1958). The multilayer extension of perceptrons, *multilayer perceptrons* (MLPs), are the kind of artificial neural networks nowadays most frequently encountered in practical applications. These will be described in some detail in the sequel. Another very popular kind of multilayer networks are the two-layer *radial basis function networks* (Buhmann, 2003), which resemble multilayer perceptrons in being used for their ability to approximate very general mappings. Among the one-layer kinds of neural networks, *associative memories* are very useful due to their ability to associate high-dimensional input patterns with low-dimensional output patterns (Hassoun, 1993). All these networks are typically used as fully connected. Hence, their architecture is uniquely described with the number of neurons in each layer. For example, the architecture of a feed-forward network with one hidden layer is uniquely described with a triple $(\boldsymbol{n}_{\mathrm{I}},\boldsymbol{n}_{\mathrm{H}},\boldsymbol{n}_{\mathrm{O}})$, in which $\boldsymbol{n}_{\mathrm{I}}$, $\boldsymbol{n}_{\mathrm{H}}$, and $\boldsymbol{n}_{\mathrm{O}}$ denote the numbers of input, hidden, and output layers, respectively. Similarly, the architecture of a network with two hidden layers is described with a quadruple $(\boldsymbol{n}_{\mathrm{I}},\boldsymbol{n}_{\mathrm{H1}},\boldsymbol{n}_{\mathrm{H2}},\boldsymbol{n}_{\mathrm{O}})$, where $\boldsymbol{n}_{\mathrm{H1}}$ and $\boldsymbol{n}_{\mathrm{H2}}$ denote the numbers of hidden neurons in the first and second layer. A very important example of networks whose architecture is inherently not fully connected are the *self-organising maps* (Kohonen, 1995; also called *Kohonen nets*), which have a remarkable ability to cluster data, especially in situations where the exact number of clusters is not known in advance. Another example are *stacked neural networks* (Sridhar *et al.*, 1996), which linearly combine several fully connected multilayer perceptrons in their output layer. Such optimal linear combinations of MLPs have also been employed to compute nonlinear regression models in catalysis (Cundari *et al.*, 2001; Tompos *et al.*, 2005).

For illustration, two multilayer perceptrons are shown in Figure 6.2. The labels assigned to the input and output neurons indicate the support and the metal additives available for composing the catalyst, and considered performance measures, respectively. Both were inspired by the case study introduced in the preceding chapter. In connection with the MLPs in Figure 6.2, two remarks are appropriate:

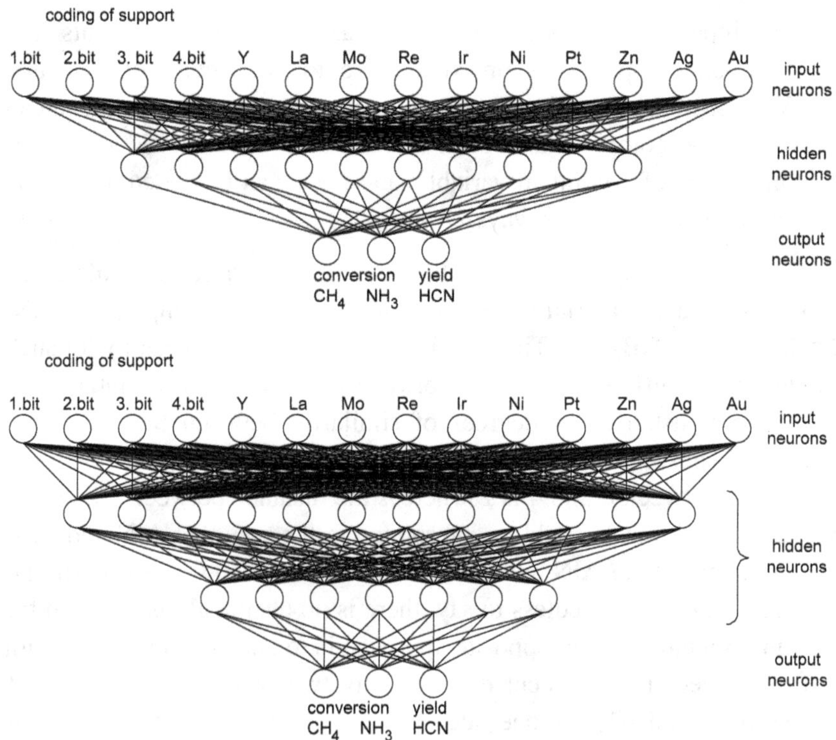

Figure 6.2. Example MLP architectures with one hidden layer (above) and two hidden layers (below), used when approximating dependencies of degree of conversion and yield on the composition of catalytic materials in the case study introduced in the preceding chapter.

(i) Although 11 compounds are available for active layers, and those compounds provide 11 metal additives, only ten additives occur among the labels of the input neurons of both networks. The reason for this is that the proportions of any ten metals already uniquely determine the proportion of the remaining 11th metal, because all proportions have to sum up to 100%. In Figure 6.2, the remaining 11th metal is Zr — the least frequent one in the data gathered in the case study.

(ii) The dependent variables can be directly assigned to ANN outputs; each neural network considered therefore needs to have *three output neurons*. Unfortunately, the situation with the independent variables and input neurons is not quite so straightforward. The inputs to a function computed by an MLP have to be *vectors of numbers*. Whereas the proportions of metal additives are numbers, and so can be used as components of such vectors, the materials used as support are values of a nominal variable, and these can be coded as vectors of numbers in various ways.

One simple way is to code all materials with a single scalar, i.e., to assign to each material a particular number — for example, 1 ≈ AlN, 2 ≈ Mo_2C, 3 ≈ TiB_2, 4 ≈ TiN, ... However, a serious problem with such coding is that different distances between numbers could be interpreted as different distances or degrees of similarity between the individual materials. Thus, in the above example, the fact that the distance between 1 and 3 or between 2 and 4 is 2 whereas the distance between 1 and 2 or between 2 and 3 is 1, could be interpreted in the way that AlN and TiB_2 are less similar than AlN and Mo_2C, and TiN and Mo_2C are less similar than TiN and TiB_2. Needless to say, there is no chemical justification for such interpretations. The opposite approach is to code each material with a separate vector component that has only two possible values (e.g., 1 and 0, or 1 and −1) — one such value indicating that the particular material is used as support in the considered catalyst, the other indicating that it is not used as support. Hence, the support is always coded with 15 those two-valued components, from which always one has the first value, the other 14 having the second value. Such coding substantially increases the number of input neurons. Observe, however, that to prevent unjustified comparisons occurring in the case of scalar coding, it is sufficient to restrict all the vector components that code the support materials to two-valued components, i.e., to components containing only *one bit of information*. Since with $\boldsymbol{c}$ bits of information, altogether 2^c different values can be coded, the 15 different materials can be coded with four bits of information in two-valued components. This is the way supports were coded for data from the case study. The precise codes for the individual materials are given in Table 6.1. Each neural network considered then needs to have *14 input neurons*: four of them coding the material used as support, the remaining ten corresponding to the proportions of the ten metal additives.

Table 6.1 Coding for the materials used as support.

Material	AlN	Mo_2C	TiB_2	TiN	NB_2O_3	BN	ZrO_2	Sm_2O_3	SrO	CaO	MgO	TiO_2	alsint	SiC	Si_3N_4
1. bit	-1	-1	-1	-1	-1	-1	-1	1	1	1	1	1	1	1	1
2. bit	-1	-1	-1	1	1	1	1	-1	-1	-1	-1	1	1	1	1
3. bit	-1	1	1	-1	-1	1	1	-1	-1	1	1	-1	-1	1	1
4. bit	1	-1	1	-1	1	-1	1	-1	1	-1	1	-1	1	-1	1

6.1.3. *Activity of Neurons*

Besides neurons and connections, another very basic concept pertaining to artificial neural networks is the *activity of a neuron.* This concept serves as a way of describing the propagation of signals from the input neurons to the output neurons. The activity of an input neuron can be thought of as coding, in terms of some appropriate units, the intensity of the signal that the neuron currently receives. Similarly, the activity of an output neuron codes the intensity of the signal that the neuron currently sends out. On the other hand, the activities of hidden neurons cannot be directly observed in reality — it is due to this property that they are called "hidden". In accordance with biological neural networks, the activity of a neuron $\boldsymbol{v}$ is supposed to be time-dependent. Consequently, this can generally be represented as a real function z_v on some set $\boldsymbol{T}$ representing time ($\boldsymbol{T}$ can be, for example, the set of non-negative real numbers, the set of non-negative integers, or some bounded interval of real numbers or integers). Very often, z_v is required to have a restricted range, typically [0,1] or [-1,1]. The system $z(t) = (z_v(t))_{v \in V}$ of activity values at time $\boldsymbol{t}$ is often called the *state* of the ANN at $\boldsymbol{t}$.

In connection with the propagation of signals from the input neurons to the output neurons, two important aspects must be taken into consideration:

- Whether the delay of signals due to the finite speed of their transmission must be taken into account. In the kinds of artificial neural networks in which that delay is considered, it always has a crucial importance. Examples include *recurrent neural networks* (Hammer, 2000), *time-delay neural networks* (Clouse *et al.*, 1997), *Hopfield nets* (Coughlinn and Baran, 1995). Various kinds of networks based on the *adaptive resonance theory* (Carpenter and Grossberg, 1987; Williamson, 1996) also belong to this group. However, networks of these types are definitely not among those most widely encountered in applications. For the purposes of this monograph, the final velocity of signal transmission can be neglected.

- Whether the network is in a steady state, in which particular signals received by the input neurons always lead to the same signals sent by the output neurons, or whether it evolves over time, allowing a particular input signal to be answered each time by another output signal. Although neural networks in steady states do have considerable importance (Section 6.2 actually deals with such networks), it is the capacity to evolve that is the most distinguishing feature of artificial neural networks; this feature allows them to adapt to the received signals as well as to an outside feedback. Due to the adaptability, the evolution of neural networks is usually called *learning* or *training*. Though several kinds of ANN training exist (supervised/unsupervised, discrete/continuous), for the purposes of this monograph we need to understand only one of them, called *supervised discrete traning*, which will be explained in Section 6.3.

6.1.4. *What Do Neural Networks Compute?*

In a network free from delays, signal transmission is simply a mapping of the received signals to the sent signals — in other words — a mapping of activities of the input neurons to activities of the output neurons. Such mapping is usually said to be *computed by the network.* If an ANN has $\boldsymbol{n_I}$ input neurons and $\boldsymbol{n_O}$ output neurons, then each mapping $\boldsymbol{F}$ computed by the network maps some subset of the $\boldsymbol{n_I}$-dimensional space into the $\boldsymbol{n_O}$-dimensional space. However, the network cannot compute an arbitrary mapping of that kind, but only a mapping $\boldsymbol{F}$ that reflects the architecture of the neural network in the sense that:

(i) The mapping $\boldsymbol{F}$ is composed gradually from a number of simple mappings, each of which is coupled either with a particular hidden or output neuron or with a particular connection between neurons. Mapping in the first case is referred to as *somatic mapping*, and mapping in the second case as *synaptic mapping* (due to the use of the terms "soma" and "synapsis" for a biological neuron and a connection between neurons, respectively). Each somatic and each synaptic mapping may only be chosen within prescribed particular simple mappings. Moreover, the same set of simple mappings is typically used for all synaptic mappings composing $\boldsymbol{F}$, whereas in the case of somatic mappings, differences may eventually exist only between mappings coupled with hidden neurons and those coupled with output neurons. Examples will be given in the next section (see also Figure 6.3 and Figure 6.4).

(ii) The way in which $\boldsymbol{F}$ is composed from somatic and synaptic mappings precisely reflects the structure of the network connections.

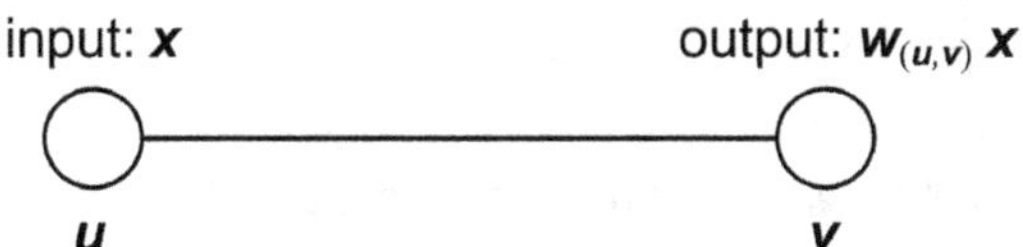

Figure 6.3. Synaptic mapping coupled with a connection (u,v) of a multilayer perceptron.

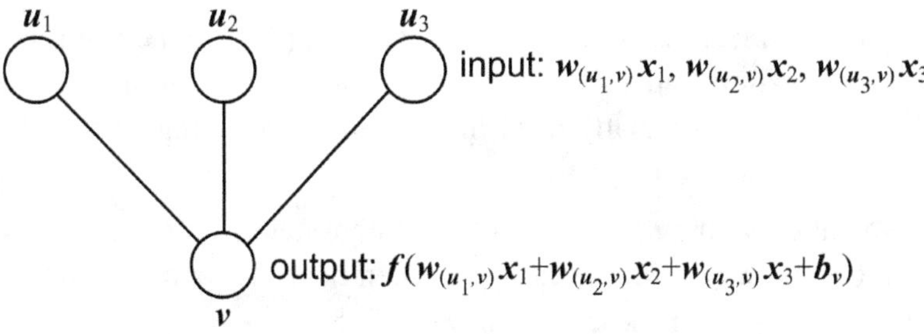

Figure 6.4. Somatic mapping coupled with a hidden neuron v of a multilayer preceptron.

For further information about basic concepts pertaining to artificial neural networks in general, and to multilayer perceptrons in particular, the reader is referred to specialised monographs such as White (1992), Hagan *et al.* (1996), Mehrotra *et al.* (1996) and Haykin (1999). An overview of traditional kinds of ANN applications in chemistry can most easily be obtained from the books by Zupan and Gasteiger (1993, 1999), and from the survey papers by Melssen *et al.* (1994), Smits *et al.* (1994) and Henson (1998).

6.2. Approximation Capability of Neural Networks

As mentioned in the preceding section, multilayer perceptrons are the kind of ANNs most frequently used in practical applications. A multilayer perceptron can be distinguished not only by its multilayer architecture with at least one hidden layer, but also by the following features:

1. Synaptic mappings are constant multiplications — that is, the synaptic mapping coupled with any connection (u,v) is the multiplication of the input value in u with some *weight* $w_{(u,v)}$ assigned to (u,v).
2. A somatic mapping coupled with a hidden neuron is chosen from among mappings that are composed of:
   (i) The summation of the results of the synaptic operations coupled with the incoming connections, and of some constant b, called *bias*.
   (ii) A nonlinear function of one variable, applied to the result of that summation and called *activation function*. The same activation function f is usually used in all somatic mappings coupled with hidden neurons.
3. A somatic mapping coupled with an output neuron is either composed in the same way as in the case of hidden neurons, or it is chosen from among summations described in 2(i).
4. Usually, there is a restriction on the nonlinearity of the activation function f, most often the restriction that f should be non-decreasing

and bounded both from below and from above, typically through the bounds 0 and 1, or -1 and 1. Functions with such a nonlinearity are called *sigmoidal functions*. Their most common examples among continuous functions are hyperpolic tangent, arctangent, and the *logistic activation function*,

$$f(x) = \frac{1}{1+e^{-x}}$$

as well as linear transformations of them; among non-continuous functions the *step function*

$$f(x) = \begin{cases} 1 & \text{if } x \geq 0 \\ 0 & \text{if } x < 0 \end{cases}.$$

Taking into account the features 1–3 above, an MLP with one hidden layer, an architecture (n_I, n_H, n_O) and an activation function f assigned only to hidden neurons computes a function $F = (F_1, \ldots, F_{n_O})$ such that for each n_I-dimensional input vector x, the scalar function F_i returns

$$F_i = w_{O,i}^T \begin{pmatrix} f\left(w_{H,1}^T x + b_{H,1}\right) \\ \cdots \\ f\left(w_{H,n_H}^T x + b_{H,n_H}\right) \end{pmatrix} + b_{O,i}. \tag{2}$$

Here, $w_{O,i}$ is an n_H-dimensional vector of weights of connections between neurons of the hidden layer and the i-th neuron of the output layer, and $b_{O,i}$ is the bias of the i-th neuron of the output layer, whereas $w_{H,j}$ is an n_I-dimensional vector of weights of connections between neurons of the input layer and the j-th neuron of the hidden layer, $b_{H,j}$ is the bias of the j-th neuron of the hidden layer, and w^T stands for the transpose of a vector w. Similarly, an MLP with two hidden layers, an architecture $(n_I, n_{H1}, n_{H2}, n_O)$ and an activation function f assigned only to hidden neurons computes a function $F = (F_1, \ldots, F_{n_O})$ such that for an input vector x, the function F_i returns

$$F_i = w_{O,i}^{\mathsf{T}} \begin{pmatrix} f\left(w_{H2,1}^{\mathsf{T}} \begin{pmatrix} f\left(w_{H1,1}^{\mathsf{T}} x + b_{H1,1}\right) \\ \cdots \\ f\left(w_{H1,n_{H1}}^{\mathsf{T}} x + b_{H1,n_{H1}}\right) \end{pmatrix} + b_{H2,1} \right) \\ \cdots \\ f\left(w_{H2,n_{H2}}^{\mathsf{T}} \begin{pmatrix} f\left(w_{H1,1}^{\mathsf{T}} x + b_{H1,1}\right) \\ \cdots \\ f\left(w_{H1,n_{H1}}^{\mathsf{T}} x + b_{H1,n_{H1}}\right) \end{pmatrix} + b_{H2,n_{H2}} \right) \end{pmatrix} + b_{O,i}. \quad (3)$$

Here, $w_{O,i}$ is an n_{H2}-dimensional vector of weights of connections between neurons of the second hidden layer and the i-th neuron of the output layer, and $b_{O,i}$ is the bias of the i-th neuron of the output layer. Similarly, $w_{H2,j}$ is an n_{H1}-dimensional vector of weights of connections between neurons of the first hidden layer and the j-th neuron of the second hidden layer, and $b_{H2,j}$ is the bias of the j-th neuron of the second hidden layer. Finally, $w_{H1,k}$ is an n_I-dimensional vector of weights of connections between neurons of the input layer and the k-th neuron of the first hidden layer, and $b_{H1,k}$ is the bias of the k-th neuron of the first hidden layer.

Such functions account for the most useful feature of multilayer perceptrons: their *universal approximation capability*. Superficially, the universal approximation capability of MLPs (as well as of some other kinds of neural networks) means that *virtually any unknown dependence can be arbitrarily closely approximated by a function computed by some MLP*. In catalysis, there is most frequently an interest in dependencies of product yields, catalyst activity, degrees of reactant conversion, and product selectivities on the input variables, such as proportions of catalyst components and reaction conditions.

For those readers who wish to fully comprehend the meaning of "virtually any dependence can be arbitrarily closely approximated by a function computed by some MLP", three points need to be clarified:

(i) If the dependence to be approximated can be represented with a function $\boldsymbol{D}$ defined on a restricted area $\boldsymbol{S}$, the expression *virtually any dependence* corresponds to functions that have a finite integral of the absolute value with respect to at least one measure $\boldsymbol{\mu}$:

$$\int_S |D|\, d\mu < \infty . \qquad (4)$$

In applications, we always deal with restricted areas because the ranges of measurements of any kind are always restricted. The *space of functions* that fulfil (4) is in mathematics denoted $L_1(\mu)$. It is supposed that all functions that can occur in any real-world application belong to such space for at least one measure μ. Moreover, that space has several *subspaces* of functions with important specific properties:

- The space $L_q(\mu)$ of functions that have a finite integral of the q-th power of the absolute value with respect to μ.
- The space $C(S)$ of functions continuous on S.
- The space $C_q(S)$ of functions continuous on S together with partial derivatives up to the order q (i.e., together with all partial derivatives $(\partial^{\alpha_1+\ldots+\alpha_n} D)/(\partial^{\alpha_1} x_1 \ldots \partial^{\alpha_n} x_n)$ such that $\alpha_1+\ldots+\alpha_n \leq q$).

(ii) The fact that D should be *arbitrarily closely* approximated by F entails the necessity of computing some *distance* $d(F,D)$ between F and D. Such distances are computed differently in different spaces; thus, if both D and F belong to several spaces simultaneously — which is typically the case — then several different distances can be computed between them.

- In the spaces $L_q(\mu)$:

$$d_q(F,D) = \left(\int_S |F - D|^q d\mu \right)^{\frac{1}{q}},$$

- in particular in $L_1(\mu)$:

$$d_1(F,D) = \int_S |F - D|\, d\mu .$$

Notice that the value $d_q(D,F)$ (in particular, $d_1(D,F)$) may be very small even if $F(x)$ and $D(x)$ substantially differ for some points x from the domain S. Hence, the fact that F is very close to the unknown dependence D in some space $L_q(\mu)$ (in particular, in $L_1(\mu)$) does not imply that the value $F(x)$ is also close to $D(x)$ for every x.

- In the space $\boldsymbol{C}(\boldsymbol{S})$:

$$d^C(\boldsymbol{F},\boldsymbol{D}) = \max_{x \in S} |\boldsymbol{F}(\boldsymbol{x}) - \boldsymbol{D}(\boldsymbol{x})|.$$

Hence, the closeness of $\boldsymbol{F}$ and $\boldsymbol{D}$ in the space $\boldsymbol{C}(S)$ also implies the closeness of $\boldsymbol{F}(\boldsymbol{x})$ and $\boldsymbol{D}(\boldsymbol{x})$ in any input point $\boldsymbol{x}$.

- Finally, in the space $\boldsymbol{C}_q(\boldsymbol{S})$:

$$d_q^C(\boldsymbol{F},\boldsymbol{D}) = \max_{\alpha_1+\ldots+\alpha_n \le q} \max_{x \in S} \left| \frac{\partial^{\alpha_1+\ldots+\alpha_n} \boldsymbol{F}(\boldsymbol{x})}{\partial^{\alpha_1} x_1 \ldots \partial^{\alpha_n} x_n} - \frac{\partial^{\alpha_1+\ldots+\alpha_n} \boldsymbol{D}(\boldsymbol{x})}{\partial^{\alpha_1} x_1 \ldots \partial^{\alpha_n} x_n} \right|.$$

The closeness of $\boldsymbol{F}$ and $\boldsymbol{D}$ in the space $\boldsymbol{C}_q(\boldsymbol{S})$ implies not only the closeness of $\boldsymbol{F}$ $(\boldsymbol{x})$ and $\boldsymbol{D}$ $(\boldsymbol{x})$, but also the closeness of their partial derivatives up to the order $\boldsymbol{q}$ for any input $\boldsymbol{x}$.

(iii) Last but not least, if the activation function is not constant and is restricted (which, in particular, covers all sigmoid activation functions), then the term *function computed by an MLP* precisely means

there exists a number $\boldsymbol{n}_{\mathrm{H}}$ of hidden neurons, as well as weights and biases $\boldsymbol{w}_{\mathrm{O},1},\boldsymbol{b}_{\mathrm{O},1},\ldots,\boldsymbol{w}_{\mathrm{O},n_\mathrm{O}},\boldsymbol{b}_{\mathrm{O},n_\mathrm{O}},\boldsymbol{w}_{\mathrm{H},1},\boldsymbol{b}_{\mathrm{H},1},\ldots,\boldsymbol{w}_{\mathrm{H},n_\mathrm{H}},\boldsymbol{b}_{\mathrm{H},n_\mathrm{H}}$, such that the computed function $\boldsymbol{F} = (\boldsymbol{F}_1,\ldots,\boldsymbol{F}_{n_\mathrm{O}})$ is defined by (2). (5)

The above formulation was adapted from Hornik (1991). However, also other results concerning approximation by means of feed-forward neural networks (Kůrková, 1992; Hornik *et al.*, 1994; Pinkus, 1998; Kůrková, 2002; Kainen *et al.*, 2007) rely on essentially the same paradigm: the required number of hidden neurons $\boldsymbol{n}_{\mathrm{H}}$ is unknown; the only guarantee is that some $\boldsymbol{n}_{\mathrm{H}}$ always exists such that a multilayer perceptron with $\boldsymbol{n}_{\mathrm{H}}$ hidden neurons can compute a function $\boldsymbol{F}$ with desired properties. The actual value of $\boldsymbol{n}_{\mathrm{H}}$ depends on the approximated dependence $\boldsymbol{D}$ and on the aspects discussed in the previous points — the function space considered and the required degree of closeness between $\boldsymbol{F}$ and $\boldsymbol{D}$; however, the fact that such an $\boldsymbol{n}_{\mathrm{H}}$ exists is independent of $\boldsymbol{D}$ and of those aspects.

6.3. Training Neural Networks

If we denote as $\mathcal{F}$ the set of all mappings from the input space of $\boldsymbol{n_I}$-dimensional vectors into the output space of $\boldsymbol{n_O}$-dimensional vectors that fulfil the restrictions (i) and (ii) listed in Subsection 6.1.4., then training neural networks without delay can be considered as nothing else than time evolution of computed mapping within the set $\mathcal{F}$. In the case of *supervised discrete learning*, that evolution has several specific features:

- The term *discrete learning* means that the evolution of the computed mapping takes place at a sequence of separate time instants, at which the network computes a sequence of mappings $\boldsymbol{F}^{(0)},\boldsymbol{F}^{(1)},\boldsymbol{F}^{(2)},\ldots$
- The term *supervised learning* means that an outside feedback is available in the form of a finite number of pairs $(\boldsymbol{x}_1,\boldsymbol{y}_1),(\boldsymbol{x}_2,\boldsymbol{y}_2),\ldots,(\boldsymbol{x}_p,\boldsymbol{y}_p)$, called *training pairs* or *training data,* such that $\boldsymbol{x}_j$ for $\boldsymbol{j} = 1,\ldots,\boldsymbol{p}$, is an $\boldsymbol{n_I}$-dimensional vector and $\boldsymbol{y}_j$ is an $\boldsymbol{n_O}$-dimensional vector of desired activities of the output neurons provided the activities of the input neurons form the vector $\boldsymbol{x}_j$. Since the activities of the input and output neurons are coding, respectively, the received and sent signals, we can also say that $\boldsymbol{y}_j$ is the vector of signals desired to be sent provided that the vector of signals $\boldsymbol{x}_j$ is received. Recall that the actual activities of the output neurons when the activities of the input neurons are $\boldsymbol{x}_j$, i.e., the signals actually sent when the vector of signals $\boldsymbol{x}_j$ is received, form the vector $\boldsymbol{F}(\boldsymbol{x}_j)$.
- In mathematical terms, the objective of supervised discrete ANN learning is to evolve the sequence $\boldsymbol{F}^{(0)},\boldsymbol{F}^{(1)},\boldsymbol{F}^{(2)},\ldots$ of computed mappings in such a way that for each $\boldsymbol{j} = 1,\ldots,\boldsymbol{p}$ the error of the vector $\boldsymbol{F}(\boldsymbol{x}_j)$ of the actual activities of the output neurons with respect to the vector $\boldsymbol{y}_j$ of the desired activities decreases. Suppose we can quantify that error with an expression $\boldsymbol{E}(\boldsymbol{y}_j,\boldsymbol{F}(\boldsymbol{x}_j))$ where $\boldsymbol{E}$ is a non-negative function on pairs of $\boldsymbol{n_O}$-dimensional vectors, i.e., a function of $2\boldsymbol{n_O}$ variables. Then, ideally, the sequence $\boldsymbol{F}^{(0)},\boldsymbol{F}^{(1)},\boldsymbol{F}^{(2)},\ldots$ should evolve according to the condition:

  Decease the sequence of errors
  $\boldsymbol{E}(\boldsymbol{y}_j,\boldsymbol{F}^{(0)}(\boldsymbol{x}_j)),\boldsymbol{E}(\boldsymbol{y}_j,\boldsymbol{F}^{(1)}(\boldsymbol{x}_j)),\boldsymbol{E}(\boldsymbol{y}_j,\boldsymbol{F}^{(2)}(\boldsymbol{x}_j)),\ldots$
  for each j = 1,…,p.

Unfortunately, if the sequence $\boldsymbol{E(y_j,F^{(0)}(x_j))}$, $\boldsymbol{E(y_j,F^{(1)}(x_j))}$, $\boldsymbol{E(y_j,F^{(2)}(x_j))}$,..., for some $\boldsymbol{j} = 1,\dots,p$ decreases close to its minimum, then the sequence $\boldsymbol{E(y_{j'},F^{(0)}(x_{j'}))}$, $\boldsymbol{E(y_{j'},F^{(1)}(x_{j'}))}$, $\boldsymbol{E(y_{j'},F^{(2)}(x_{j'}))}$,... must necessarily increase for at least one $\boldsymbol{j'}$ different from j, unless there exists a mapping $\boldsymbol{F}$ in the set $\mathcal{F}$ that exactly fits all the training pairs $(x_1,y_1),(x_2,y_2),\dots,(x_p,y_p)$. And exactly fitting all available training data is usually connected with the inability to capture general dependencies reflected in the training pairs (this phenomenon, referred to as overtraining, will be described later). That is the reason why instead of the above condition, the sequence $\boldsymbol{F^{(0)},F^{(1)},F^{(2)}}$,... is generally evolved according to the following weaker condition:

Decrease a sequence of weighted combinations of errors
$c_1E(y_1,F^{(0)}(x_1))+c_2E(y_2,F^{(0)}(x_2))+\dots+c_pE((y_p,F^{(0)}((x_p)),$
$c_1E(y_1,F^{(1)}(x_1))+c_2E(y_2,F^{(1)}(x_2))+\dots+c_pE((y_p,F^{(1)}((x_p)),$
$c_1E(y_1,F^{(2)}(x_1))+c_2E(y_2,F^{(2)}(x_2))+\dots+c_pE((y_p,F^{(2)}((x_p)),\dots,$

where $c_1,c_2,\dots,c_p$ are positive constants, indicating a relative importance of the individual training pairs $(x_1,y_1),(x_2,y_2),\dots,(x_p,y_p)$, respectively, for the evolution of the sequence $\boldsymbol{F^{(0)},F^{(1)},F^{(2)}}$,.... In particular, if all training pairs are considered to be equally important, then the weighted combination can be set to their average:

Decrease the sequence of average errors
$[(E(y_1,F^{(0)}(x_1))+\dots E((y_p,F^{(0)}((x_p)),E(y_1,F^{(1)}(x_1))+\dots$
$\dots E((y_p,F^{(1)}((x_p)),E(y_1,F^{(2)}(x_1))+\dots E((y_p,F^{(2)}((x_p)),\dots)]\,/\,p.$

- Learning ends in some step $\boldsymbol{s}$ — i.e., after the evolution through a sequence of computed mappings $\boldsymbol{F^{(0)},F^{(1)},\dots,F^{(s)}}$, either due to fulfilling some prescribed stopping criterion, or through a deliberate break from outside. A stopping criterion is typically some condition, or a combination of conditions, on:

  - the step $\boldsymbol{s}$,
  - the considered weighted combination of errors $c_1E(y_1,F^{(s)}(x_1))+\dots+$ $c_2E(y_2,F^{(s)}(x_2))+\dots+c_pE((y_p,F^{(s)}((x_p)),$

- the difference of such combinations between the current and the previous step, $c_1[E(y_1,F^{(s-1)}(x_1)) - E(y_1,F^{(s)}(x_1))] + c_2[E(y_2,F^{(s-1)}(x_2)) - E(y_2,F^{(s)}(x_2))] + \ldots + c_p[E((y_p,F^{(s-1)}((x_p)) - E((y_p,F^{(s)}((x_p))]$,
- the size of the gradient of the considered weighted combination of errors with respect to parameters that parametrise the computed mappings, i.e., $c_1\nabla E(y_1,F^{(s)}(x_1)) + \ldots + c_2\nabla E(y_2,F^{(s)}(x_2)) + \ldots + c_p\nabla E((y_p,F^{(s)}((x_p))$, provided E is differentiable.

The above description of learning indicates the crucial importance of the errors of the vector $F(x_j)$ of the actual activities of the output neurons and the vector y_j of their desired activities, captured by means of a nonnegative function E on pairs of vectors in the n_O-dimensional output space. Quite a large number of such functions has already been proposed, due to the fact that most of them are applicable only in specific situations. Nevertheless, two of them are universally applicable. The most frequently encountered error function is based simply on the usual (i.e., Euclidean) distance in the output space, i.e. on the distance induced by the Euclidean norm $\|\ \|$. From the computational point of view, however, more advantageous than the Euclidean norm itself is its square, since this can be quickly computed as the sum of squares of individual components. Therefore, this most common error function is called *sum of squared errors* (SSE):

$$\mathrm{SSE}(\hat{y},y) = \|\hat{y} - y\|^2 = \sum_{i=1}^{n_O} (\hat{y}_i - y_i)^2 ,$$

where $\hat{y}$ is the vector of the actual activities of the input neurons, y is the vector of their desired activities, and $\hat{y}^i, y^i$ for $i = 1,\ldots,n_O$ denote their i-th coordinates. Recall that if all training pairs $(x_1,y_1),(x_2,y_2),\ldots,(x_p,y_p)$ used for supervised learning are considered equally important, then the condition determining how the sequence $F^{(0)},F^{(1)},F^{(2)},\ldots$ evolves is to decrease the sequence of averaged errors. That sequence is in the case of the function SSE called the *mean squared error* (MSE) of p pairs of vectors in the n_O-dimensional output space:

$$\mathrm{MSE}\left((\hat{y}_1,y_1),\ldots,(\hat{y}_p,y_p)\right) = \frac{1}{p}\sum_{j=1}^{p} \|\hat{y}_j - y_j\|^2 = \frac{1}{p}\sum_{i=1}^{n_O}\sum_{j=1}^{p} (\hat{y}_j^i - y_j^i)^2 . \quad (6)$$

Needless to say, the usual Euclidean norm is not the only norm in the n_O-dimensional output space. In particular, replacing the squares in the definition of SSE with a general q-th power for $q \geq 1$ leads to the error function sum of q-th powers of errors (SPE):

$$\mathrm{SPE}(\hat{y}, y) = \sum_{i=1}^{n_O} \left(\hat{y}^i - y^i\right)^q ,$$

which is, like the disagreement SSE, the q-th power of a norm on the n_O-dimensional output space. If the activities of the output neurons can be interpreted as a probability distribution on some finite set, which is usually the case if the network is used to perform classification of inputs into n_O classes, then in addition to the sum of squares, the information-theoretical error function called *relative entropy* (RE) is frequently used:

$$\mathrm{RE}(\hat{y}, y) = \sum_{i=1}^{n_O} \left[y_i \log \frac{y^i}{\hat{y}^i} + \left(1 - y^i\right) \log \frac{1 - y^i}{1 - \hat{y}^i} \right],$$

where the n_O-dimensional vectors $\hat{y}$ and y are discrete probability distributions on n_O points.

Another error function is available in the frequently occurring situation that the desired probability distribution y is degenerated, or equivalently that the network input should be classified into exactly one class — a function called *classification figure of merit* (CFM):

$$\mathrm{CFM}(\hat{y}, y) = \frac{\alpha}{n_O - 1} \sum_{\substack{i=1, \\ i \neq i_1}}^{n_O} \frac{1}{\beta + e^{-\gamma\left((\hat{y})_{i_1} - (\hat{y})_i\right)^2}},$$

where i_1 denotes the component of y that equals 1, and α, β, γ are constants, α and γ being positive. Finally, in the case of a network with only one output neuron and with desired activities assuming only two opposite values, say -1 and 1 (for example, if the network is required to make some yes/no decision), quick learning is possible using the *zero-one error* function (ZOE):

$$\mathrm{ZOE}(\hat{y}, y) = \max\{0, -y \operatorname{sign} \hat{y}\}.$$

In the following, attention will be restricted to the error function MSE (6), i.e., to SSE combined with equal importance of all the training pairs.

Hence, the training method has to evolve the sequence of mappings $\boldsymbol{F}^{(0)},\boldsymbol{F}^{(1)},\boldsymbol{F}^{(2)},\ldots$ in such a way that the MSE of $(\boldsymbol{x}_1,\boldsymbol{y}_1),(\boldsymbol{x}_2,\boldsymbol{y}_2),\ldots,(\boldsymbol{x}_p,\boldsymbol{y}_p)$ becomes sufficiently small. To get an MSE value as small as possible, the mean squared error has to be minimised with respect to the computed mappings — thus, according to (2)–(3), with respect to the parameters that determine those mappings — weights and biases.

Superficially, the approximation result (5) in the previous section seems to suggest that the choice of an appropriate number of hidden neurons will be straightforward: since any function computed by an ANN with a certain number of hidden neurons can, according to (2)–(3), also be computed by any ANN with additional hidden neurons (it is sufficient to set all weights of connections leading to or from those additional neurons to 0), one could simply decide to choose a number of neurons that is high enough — or even better, to choose two hidden layers with sufficiently high numbers of neurons in each of them. Any such decision, however, would completely ignore the fact that the information on which network learning relies is restricted to the training data $(\boldsymbol{x}_1,\boldsymbol{y}_1),(\boldsymbol{x}_2,\boldsymbol{y}_2),\ldots,(\boldsymbol{x}_p,\boldsymbol{y}_p)$. That restriction carries with it two dangers:

1. The computed mapping $\boldsymbol{F}$, resulting from network learning, can fit the training pairs precisely, even though the latter reflect not only the unknown dependence $\boldsymbol{D}$, but also noise due to random influences, such as measurement errors.
2. In addition to precisely fitting the training data $(\boldsymbol{x}_1,\boldsymbol{y}_1),\ldots,(\boldsymbol{x}_p,\boldsymbol{y}_p)$, the computed mapping $\boldsymbol{F}$ can assume fully irrelevant values $\boldsymbol{F}(\boldsymbol{x})$ for $\boldsymbol{x}$ different from $\boldsymbol{x}_1,\boldsymbol{x}_2,\ldots,\boldsymbol{x}_p$. Hence, instead of approximating the unknown dependence, $\boldsymbol{F}$ can actually approximate a mapping that only to some degree indicates what the training pairs used for its learning were, for example,

$$\boldsymbol{F}_i(\boldsymbol{x}) = \begin{cases} y_j^i \dfrac{\|\boldsymbol{x}-\boldsymbol{x}_j\|}{\boldsymbol{d}} & \text{if } |\boldsymbol{x}-\boldsymbol{x}_j| < \boldsymbol{d} \quad \text{for some} \\ & j = 1,\ldots,\boldsymbol{p}, i = 1,\ldots,\boldsymbol{n}_O, \\ 0 & \text{otherwise,} \end{cases} \tag{7}$$

where $\boldsymbol{d} > 0$ is sufficiently small. The phenomenon whereby a neural network learns to merely indicate what the training pairs were instead of generalising to the unknown dependence, is called *overtraining*, *overlearning*, or *overfitting*.

To approximate a function like the one above, many weights and biases are needed. Recall from (2)–(3) that the number of weights and biases increases with the number of hidden neurons. Hence, avoiding a high risk of overtraining implies a requirement to keep the number of neurons low.

Since the MSE on training data is small not only in the case of overtraining, but also in the case of a good approximation, training data alone are not sufficient to recognise overtraining. A separate set of experimentally obtained *test data* $(\boldsymbol{x}_1,\boldsymbol{y}_1),\ldots,(\boldsymbol{x}_q,\boldsymbol{y}_q)$ is needed — one that does not overlap with the training data, but is still governed by the same unknown dependence $\boldsymbol{D}$ (it is immaterial whether the test catalysts were already available during network training or only obtained later, via additional catalyst experiments). In such a case, all differences between the MSE on the training and test data will be attributable to noise and overtraining. If one assumes, in addition, that the distribution of noise in the test data is the same as in the training data, the following conclusions can be drawn from the obtained MSE values:

- An MSE on the test data that is lower than, equal to or slightly higher than that on the training data can be explained as due to noise and indicates an absence of overtraining.
- An MSE on the test data that is substantially higher than that on the training data indicates overtraining (Figure 6.5).
- If the MSE on the test data up to an iteration decreases while it after this iteration increases, then overtraining is present at least beginning with the following iteration (Figure 6.6).
- The one out of two multilayer perceptrons which better approximates the unknown dependence (or the one out of the set of MLPs which approximates that dependence best) is indicated by the corresponding MSE values on test data, not by the values on training data.

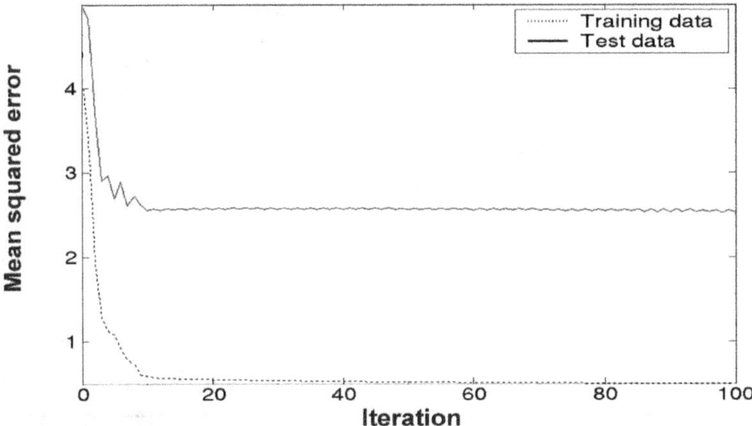

Figure 6.5. Time development of the MSE values during an example MLP training with the basic Levenberg-Marquardt method. The fact that the MSE value for the test data has a constant difference from the MSE value for the training data already from the early iterations, indicates constant overtraining from the beginning.

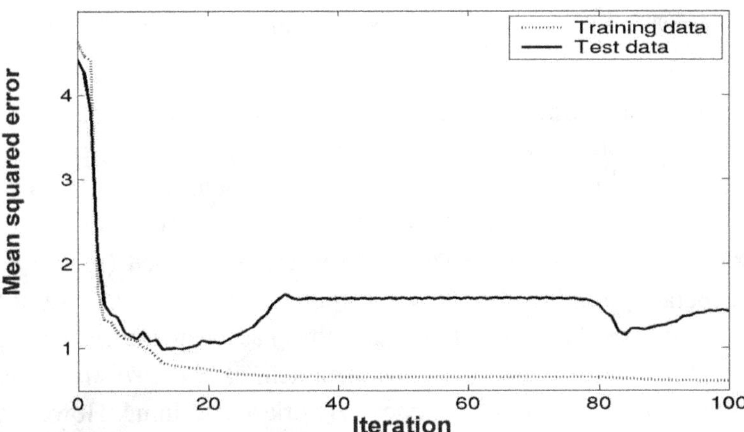

Figure 6.6. Time development of the MSE values during an example MLP training with the basic Levenberg-Marquardt method. This indicates increasing overtraining with a later onset.

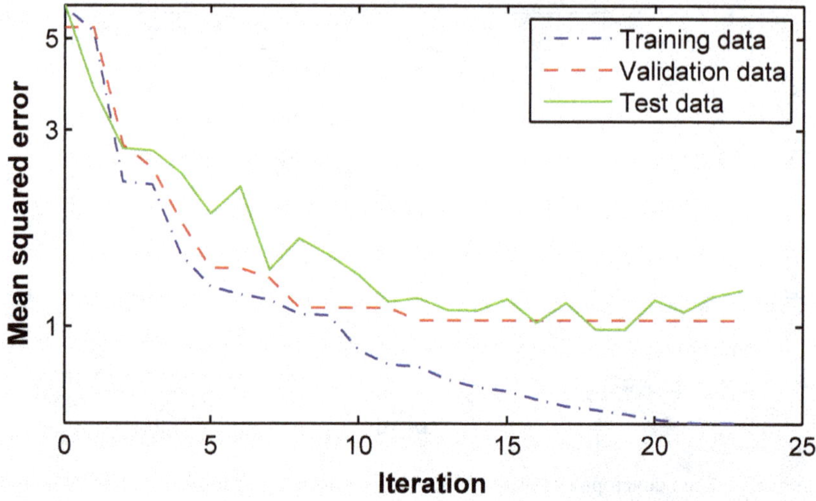

Figure 6.7. Time development of the MSE values during an example MLP training with the early-stopping variant of the Levenberg-Marquardt method. The time development ends with the 23rd iteration, since an increase in overtraining was detected starting at the 24th iteration. That detection was based on the time development of the MSE values for the validation data, and is also in accordance with the development of the MSE values for the test data, which also indicates a slight increase in overtraining from approximately the 20th iteration onward.

The conclusion drawn above that overtraining is present after the first increase of the MSE on the test data, suggests an idea as to how overtraining could be reduced: to stop network training as soon as such an increase occurs. As the final weights and biases, then, those from the last iteration before the MSE started to increase are used (Figure 6.7). More generally, training can be stopped not after the first increase of the MSE on the test data, but after that value has been increasing for a prescribed number of iterations. This approach, called *early stopping*, is indeed frequently employed to reduce network overtraining. However, if test data are used to detect overtraining, they cannot serve as test data any more because, through the detection of overtraining, they have already been involved in training. Consequently, we need two separate sets of test data for early stopping. Data for detection of overtraining are usually called *validation data*, whereas the proper test data are used to provide an overall indication of how well the network has been trained.

A more sophisticated overtraining suppression method, called *Bayesian regularisation*, relies on the viewpoint of Bayesian statistics, that weights and biases of a multilayer perceptron are actually random variables with a particular probability distribution, and that overtraining corresponds to situations where the values of those parameters obtained through MSE minimisation are unlikely with respect to that distribution (MacKay, 1992). In such a case, values of weights and biases corresponding to minimal overtraining can be obtained via replacing MSE minimisation with an MSE-based statistical parameter estimation.

6.4. Knowledge Obtainable from a Trained Network

The architecture of a trained neural network and the weights and biases that determine the mapping computed by the network inherently represent the knowledge contained in the data used to train the network. This is the knowledge about various relationships between variables corresponding to the inputs and to the outputs of the network, such as relationships between the composition of the catalytic materials and their catalytic performance in the experiments in which the data were gathered. However, such a representation is not really comprehensible for human beings, since it is very far from the symbolic, modular and vague way people represent knowledge. Hence, the larger the amounts of empirical data used to train artificial neural networks, the more important is the application of appropriate *knowledge extraction methods*. The latter term denotes methods that enable to transform inherent knowledge representation by means of network architecture and parameters determining the computed mapping into formal rules which are considered to be human-comprehensible.

Most frequently, the formal rules extracted from artificial neural networks have the form:

IF the input variables fulfil an input condition $\boldsymbol{C_I}$,
THEN the output variables are likely to fulfil an output condition $\boldsymbol{C_O}$. (8)

Over the last two decades, various rule-extraction methods have been proposed for trained neural networks, but so far none of them has become commonly used (cf. the survey papers by Andrews *et al.*, 1995; Tickle *et al.*, 1998; Mitra and Hayashi, 2000, and the monograph by Garcez *et al.*, 2009). Here, a method will be sketched that finds for each output condition of the form:

C_O: the value $\boldsymbol{y}$ of the output variables lies in a rectangular area $\boldsymbol{R}$

one or more input conditions of the form:

C_I: the value $\boldsymbol{x}$ of the input variables lies in a polyhedron $\boldsymbol{P}$.

Hence, this method extracts rules of the form:

IF $\boldsymbol{x} \in \boldsymbol{P}$, THEN $\boldsymbol{y} \in \boldsymbol{R}$.

A detailed explanation of the method can be found in Holeňa (2006). Its main principles can be summarised as follows:

- A rectangular area $\boldsymbol{R}$ with borders perpendicular to axes, defining the consequence (the THEN… part) of a rule to extract, has to be chosen in advance in the output space of a trained multilayer perceptron. Typically, this is specified by means of an inequality constraint on some of the output variables, e.g., $y_1 > 9\%$, or by means of a combination of such constraints, e.g., $y_1 > 8\%$ & $y_2 > 1.5\%$. For illustration, assume that an MLP has two output neurons corresponding to the output variables y_1 and y_2 such that y_1 records the yield of propene in an oxidative dehydrogenation of propane to propene, and y_2 records the yield of ethylene in the reaction. Then the former condition reads

  propene yield > 9%,

  whereas the latter combination reads

  propene yield > 8% and ethylene yield > 1.5%.
- The sigmoidal activation functions used in the somatic mappings coupled with hidden neurons are approximated with piecewise-linear sigmoidal activation functions. This can be done with an arbitrary precision.
- The products of individual linearity intervals of all the activation functions determine areas in the input space in which the final approximating mapping computed by the multilayer perceptron is linear.

- In each such area, all points mapped to $\boldsymbol{R}$ form a polyhedron, which may eventually be empty or may be concatenated with polyhedra from some of the neighbouring areas to a larger polyhedron.
- The union of all the nonempty concatenated polyhedra $\boldsymbol{P}_1,\ldots,\boldsymbol{P}_q$ defines the antecedent (the IF … part) of a rule in a combined form

$$\text{IF } \boldsymbol{x} \in \boldsymbol{P}_1 \cup \ldots \cup \boldsymbol{P}_q, \text{ THEN } \boldsymbol{y} \in \boldsymbol{R} \tag{9}$$

which is equivalent to a logical disjunction of $\boldsymbol{q}$ rules of the simple form (8)

IF $\boldsymbol{x} \in \boldsymbol{P}_1$, THEN $\boldsymbol{y} \in \boldsymbol{R}$

…

IF $\boldsymbol{x} \in \boldsymbol{P}_q$, THEN $\boldsymbol{y} \in \boldsymbol{R}$.

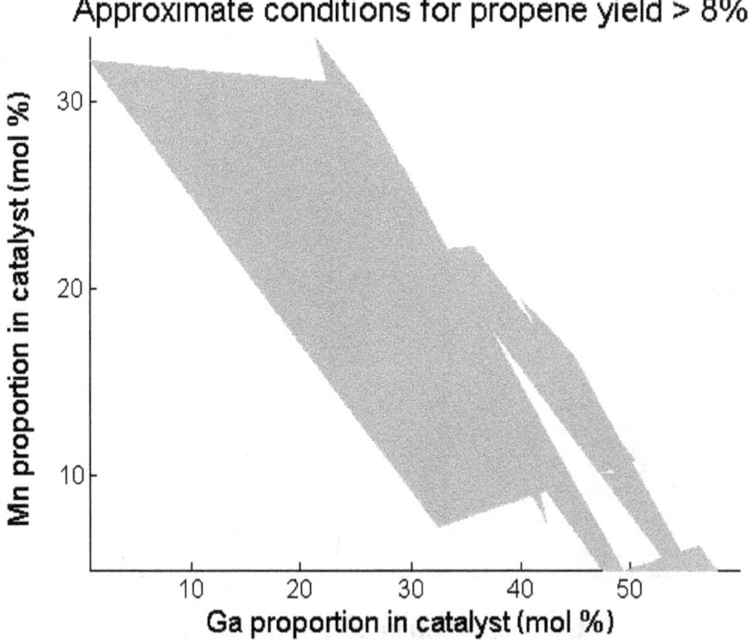

Figure 6.8. A two-dimensional cut of the union of polyhedra from the antecedent of a rule of the form (9) extracted from a trained MLP. The cut corresponds to input variables recording the molar proportions of oxides of Mn and Ga in the catalytic material for the consequence "propene yield > 8%".

To increase the comprehensibility of the extracted rules, visualisation by means of two- or three-dimensional cuts of the set $\boldsymbol{P}_1 \cup \ldots \cup \boldsymbol{P}_q$ can be used. To this end, the values of some of the input variables have to be fixed, so that the number of free input variables is restricted to two or three (the dimensionality of the cut). The resulting visualisation then depends not only on the set $\boldsymbol{P}_1 \cup \ldots \cup \boldsymbol{P}_q$, but also on the fixed values.

Examples of two-dimensional cuts visualising the unions of polyhedra that determine the antecedents of two combined-form rules extracted from the same trained MLP are shown in Figure 6.8 and Figure 6.9. Both cuts correspond to identical pairs of input variables, but to the increasingly restrictive output conditions.

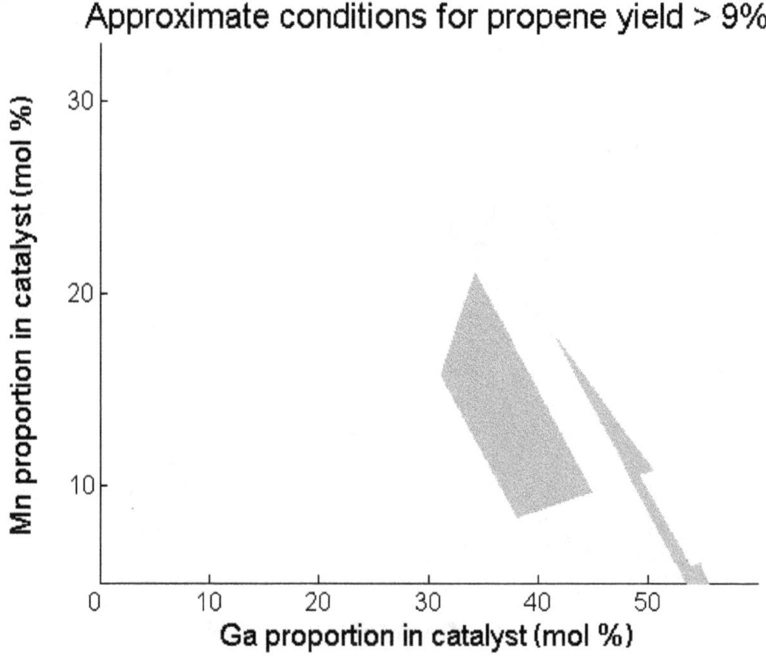

Figure 6.9. A two-dimensional cut of the union of polyhedra from the antecedent of a combined-form rule extracted from the same MLP as in Figure 6.8. This cut corresponds to the same input variables as the cut in Figure 6.8, but for a more restrictive consequence: “propene yield > 9%”.

Usually, logical rules of the above form

IF $\boldsymbol{x} \in \boldsymbol{P}$, THEN $\boldsymbol{y} \in \boldsymbol{R}$

where $\boldsymbol{P}$ is a polyhedron and $\boldsymbol{R}$ is a rectangular area, are the final results of this rule-extraction method. However, there is one exception — the case when $\boldsymbol{P}$ is also rectangular with borders perpendicular to axes, or more generally, when $\boldsymbol{P}$ can be approximately replaced with such a rectangular area $\boldsymbol{R_I}$ in the input space. Then the above rule can be approximately expressed in the conjunctive form

IF $\boldsymbol{x}_1 \in \boldsymbol{I}_1$ & $\boldsymbol{x}_2 \in \boldsymbol{I}_2$ &...& $\boldsymbol{x}_{nI} \in \boldsymbol{I}_{nI}$, THEN $\boldsymbol{y} \in \boldsymbol{R}$.

Here, $\boldsymbol{I}_1, \boldsymbol{I}_2, \ldots, \boldsymbol{I}_{nI}$ are intervals that constitute the projections of $\boldsymbol{R_I}$ into the $\boldsymbol{n_I}$ input dimensions. For example

IF $\boldsymbol{x}_1 = 0\%$ **&** $15.8\% < \boldsymbol{x}_2 < 19.3\%$ **&** $37.3\% < \boldsymbol{x}_3 < 41.2\%$ **&** $\boldsymbol{x}_4 = 0\%$ **&** $9.7\% < \boldsymbol{x}_5 < 11.9\%$ THEN $\boldsymbol{y}_1 > 8\%$.

For illustration, assume that an MLP has two output neurons corresponding to the above considered variables $\boldsymbol{y}_1$ and $\boldsymbol{y}_2$, and five input neurons corresponding to the variables $\boldsymbol{x}_1$–$\boldsymbol{x}_5$ that record, in turn, the molar proportions of the oxides of Fe, Ga, Mg, Mn and Mo in the catalytic material (more precisely, in its active shell). Then the above rule reads

IF Fe proportion in catalyst = 0%
and 15.8% < Ga proportion in catalyst < 19.3%
and 37.3% < Mg proportion in catalyst < 41.2%
and Mn proportion in catalyst = 0%
and 9.7% < Mo proportion in catalyst < 11.9%,
THEN propene yield > 8%.

Each such interval can be restricted both from below and from above, restricted only from below or only from above, or finally can be even the complete set of real numbers. However, dimensions for which the corresponding projection of $\boldsymbol{R_I}$ equals the complete real axis are usually not included in the conjunctive-form rules, since they would not provide any new knowledge.

In the rule-extraction method outlined above, the possibility of replacing a polyhedron $\boldsymbol{P}$ with a rectangular area $\boldsymbol{R_I}$ is assessed according to the following principles:

(i) The resulting dissatisfaction with points that either belong to $\boldsymbol{P}$ but do not belong to $\boldsymbol{R_I}$, or belong to $\boldsymbol{R_I}$ but do not belong to $\boldsymbol{P}$ (i.e., the dissatisfaction with points that belong to the symmetric difference $\boldsymbol{R_I \Delta P}$), has to remain within a prescribed tolerance $\boldsymbol{\varepsilon}$ and for $\boldsymbol{R_I}$ has to be minimal in the input space among rectangular areas of a specific kind.

(ii) The dissatisfaction with points from $\boldsymbol{R_I \Delta P}$ depends solely on those points and is increasing with respect to inclusion. Consequently, it can be measured using some monotone measure on the input space, possibly depending on $\boldsymbol{P}$.

(iii) To be eligible for replacement, $\boldsymbol{P}$ has to cover at least one point of the available data.

For (ii), the most attractive monotone measures, due to their straightforward interpretability, are:

- The joint empirical distribution of the input variables in the available data.
- The conditional empirical distribution of the input variables in the available data, conditioned by $\boldsymbol{P}$.

Conjunctive-form rules are also very convenient from the visualisation point of view — since cuts of rectangular areas coincide with the corresponding projections of those areas, the values of no variables need to be fixed. As an example, Figure 6.10 shows three-dimensional cuts determining the antecedents of conjunctive-form rules extracted from a trained MLP with six input neurons and two output neurons, assuming the above interpretation of the variables to which those neurons correspond. The rules are extracted according to the principles (i)–(iii) for the consequence "propene yield > 8%" and are listed in Table 6.2.

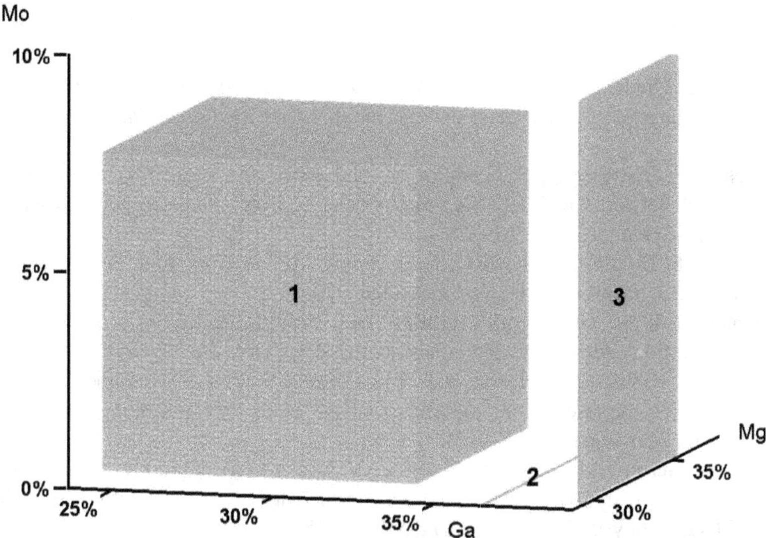

Figure 6.10. A three-dimensional cut of the union of rectangular areas that replace, in accordance with the method described in this section, the union of of polyhedra from the antecedent of a combined form rule extracted from a trained MLP. The cut corresponds to input variables recording the molar proportion of oxides of Ga, Mg and Mo in the catalytic material. The numbers 1, 2, 3 refer to the antecedent of the rules in Table 6.2.

Table 6.2. Antecedents of the conjunctive-form rules extracted from a trained MLP using the method described in this section for the consequence "propene yield > 8%".

Rule	Antecedent (IF... part)	Consequence (THEN... part)
1	24% < Ga proportion in catalyst < 33% and 31% < Mg proportion in catalyst < 39% and Mo proportion in catalyst < 7% and Fe, Mn proportions in catalyst = 0	Propene yield > 8%
2	Ga proportion in catalyst ≈ 36% and 28% < Mg proportion in catalyst < 38% and Fe, Mn, Mo proportions in catalyst = 0	
3	Fe proportion in catalyst < 12% and Ga proportion in catalyst ≈ 38% and 29% < Mg proportion in catalyst < 36% and Mo proportion in catalyst < 9% and Mn proportions in catalyst = 0	

Bibliography

Andrews, R., Diederich, J. and Tickle, A.B. (1995). Survey and critique of techniques for extracting rules from trained artificial neural networks, *Knowl. Based Syst.*, 8, 378–389.

Baumes, L.A., Farrusseng, D., Lengliz, M. and Mirodatos, C. (2004). Using artificial neural networks to boost high-throughput discovery in heterogeneous catalysis, *QSAR Comb. Sci.*, 23, 767–778.

Buhmann, M.D. (2003). *Radial Basis Functions: Theory and Implementations*, Cambridge University Press, Cambridge, 270 p.

Carpenter, G.A. and Grossberg, S. (1987). ART 2: Self-organization of stable category recognition codes for analog input patterns, *Appl. Opt.*, 26, 4919–4930.

Clouse, D.S., Giles, C.L., Horne, B.G. and Cottrell, G.W. (1997). Time-delay neural networks: representation and induction of finite-state machines, *IEEE Trans. Neural Networks*, 8, 1065–1070.

Corma, A., Serra, J.M., Argente, E., Botti, V. and Valero, S. (2002). Application of artificial neural networks to combinatorial catalysis: modeling and predicting ODHE catalysts, *ChemPhysChem*, *3*, 939–945.

Coughlin, J.P. and Baran, R.H. (1995). *Neural Computation in Hopfield Networks and Boltzmann Machines*, University of Delaware Press, Newark, 281 p.

Cundari, T., Deng, J. and Zhao, Y. (2001). Design of a propane ammoxidation catalyst using artificial neural networks and genetic algorithms, *Ind. Eng. Chem. Res.*, 40, 5475–5480.

Diaconescu, R. and Dumitriu, E. (2005). Applications of artificial neural networks in environmental catalysis, *Environ. Eng. Manage. J.*, 4, 473-498.

Farrusseng, D., Klanner, C., Baumes, L., Lengliz, M., Mirodatos, C. and Schüth, F. (2005). Design of discovery libraries for solids based on QSAR models, *QSAR Comb. Sci.*, 24, 78–93.

Garcez, A.S.A., Lamb, L.C. and Gabbay, D.M. (2009). *Neural-Symbolic Cognitive Reasoning*, Springer, Berlin, 197 p.

Gottwald, S. (1993). *Fuzzy Sets and Fuzzy Logic: The Foundations of Application — from a Mathematical Point of View,* Vieweg, Wiesbaden, 216 p.

Günay, M.E. and Yildirim, R. (2008). Neural network aided deisgn of Pt-Co-Ce/Al_2O_3 catalyst for selective CO oxidation in hydrogen-rich streams, *Chem. Eng. J.*, 140, 324–331.

Hagan, M.T., Demuth, H. and Beale, M. (1995). *Neural Network Design,* PWS Publishing, Boston, 736 p.

Hájek, P. (1998). *Metamathematics of Fuzzy Logic,*Kluwer, Dordrecht, 297 p.

Hammer, B. (2000). *Learning with Recurrent Neural Networks*, Springer, Berlin, 149 p.

Hassoun, M.H. (1993). *Associative Neural Memories: Theory and Implementation*, Oxford University Press, Oxford, 376 p.

Hattori, T. and Kito, S. (1995). Neural network as a tool for catalyst development, *Catal. Today*, 23, 347–355.

Haykin, S. (1999). *Neural Networks: A Comprehensive Foundation,* Prentice Hall of India, Delhi, 864 p.

Henson, M.A. (1998). Nonlinear model predictive control: Current status and future directions, *Comput. Chem. Eng.*, 23, 187–202.

Holeňa, M. and Baerns, M. (2003). Feedforward neural networks in catalysis: A tool for the approximation of the dependence of yield on catalyst composition, and for knowledge extraction, *Catal. Today*, 81, 485–494.

Holeňa, M. (2006). Piecewise-linear neural networks and their relationship to rule extraction from data, *Neural Comput.*, 18, 2813–2853.

Hornik, K. (1991). Approximation capabilities of multilayer neural networks, *Neural Netw.*, 4, 251–257.

Hornik, K., Stinchcombe, M., White, H. and Auer, P. (1994). Degree of approximation results for feedforward networks approximating unknown mappings and their derivatives, *Neural Comput.*, 6, 1262–1275.

Hou, Z.Y., Dai, Q., Wu, X.Q. and Chen, G.T. (1997). Artificial neural network-aided design of catalyst for propane ammoxidation. *Appl. Catal., A: General*, 161, 183–190.

Huang, K., Feng-Qiu, C. and Lü, D.W. (2001). Artificial neural network-aided design of a multi-component catalyst for methane oxidative coupling, *Appl. Catal., A: General*, 219, 61–68.

Kainen, P.C., Kůrková, V. and Sanguineti, M. (2007). Estimates of approximation rates by gaussian radial-basis functions. In: Beliczynski, B., Iwanovski, M. and Ribeiro, B. (eds.), *Adaptive and Natural Computing Algorithms*, Springer, Berlin, pp. 11–18.

Kito, S., Hattori, T. and Murakami, Y. (1994). Estimation of catalytic performance by neural network — Product distribution in oxidative dehydrogenation of ethylbenzene, *Appl. Catal., A: General*, 114, L173–L178.

Kito, S., Satsuma, A., Ishikura, T., Niwa, M., Murakami, Y. and Hattori, T. (2004). Application of neural networks to estimation in catalyst deactivation in methanol conversion, *Catal. Today*, 97, 41–46.

Klanner, C., (2004). Evaluation of descriptors for solids, Thesis, Ruhr-University Bochum, 198 p.

Kohonen, T. (1995). *Self-organizing maps*, Springer, Berlin, 362 p.

Kůrková, V. (1992). Kolmogorov's theorem and multilayer neural networks, *Neural Netw.*, 5, 501–506.

Kůrková, V. (2002). Neural networks as universal approximators. In: Arbib, M.A. (ed.), *Handbook of Brain Theory and Neural Networks*, MIT Press, Cambridge (MA), pp. 1180–1183.

Liu, Y., Liu, D., Cao, T., Han, S. snd Xu, G. (2001). Design of a CO_2 hydrogenation catalyst by an artificial neural network, *Comput. Chem. Eng.*, 25, 1711–1714.

MacKay, D.J.C. (1992). Bayesian interpolation, *Neural Comput.*, 4, 415–447.

Mehrotra, K., Mohan, C.K. and Ranka, S. (1997). *Elements of Artificial Neural Networks,* MIT Press, Cambridge (MA), 360 p.

Melssen, W.J., Smits, J.R.M., Buydens, L.M.C. and Kateman, G. (1994). Using artificial neural networks for solving chemical problems. Part II: Kohonen self-organizing feature maps and Hopfield networks, *Chemom. Intell. Lab. Systems*, 23, 267–291.

Mitra, S. and Hayashi, Y. (2000). Neuro-fuzzy rule generation: Survey in soft computing framework, *IEEE Trans. Neural Networks*, 11, 748–768.

Moliner, M., Serra, J.M., Corma, A., Argente, E., Valero, S. and Botti, V. (2005). Application of artificial neural networks to high-throughput synthesis of zeolites, *Microporous Mesoporous Mater.*, 78, 73–81.

Novák, V., Perfilieva, I. and Močkoř, J. (1999). *Mathematical Principles of Fuzzy Logic*, Kluwer, Boston, 320 p.

Omata, K., Hashimoto, M., Watanabe, Y., Umegaki, T., Wagatsuma, S., Ishiguro, G. and Yamada, M. (2004). Optimization of Cu oxide catalyst for methanol synthesis under high CO_2 partial pressure using combinatorial tools, *Appl. Catal., A: General*, 262, 207–214.

Omata, K., Kobayashi, Y. and Yamada, M. (2005). Artificial neural network-aided development of supported Co catalyst for preferential oxidation of CO in excess hydrogen, *Catal. Commun.*, 6, 563–567.

Pawlak, Z. (1991). *Rough Sets: Theoretical Aspects of Reasoning about Data*, Springer, Berlin, 252 p.

Pinkus, A. (1996). Approximation theory of the MLP model in neural networks, *Acta Numer*, 8, 277–283.

Rosenblatt, F. (1958). The perceptron: A probabilistic model for information storage and organization in the brain, *Psycholog. Rev.*, 65, 386–486.

Sasaki, M., Hamada, H., Kintaichi, Y. and Ito, T. (1995). Application of a neural network to the analysis of catalytic reactions. Analysis of NO decomposition over Cu/ZSM-5 zeolite, *Appl. Catal., A: General*, 132, 261–270.

Serra, J.M., Baumes, L.A., Moliner, M., Serna, P. and Corma, A. (2007). Zeolite synthesis modelling with support vector machines: a combinatorial approach, *Comb. Chem. High Throughput Screening*, 10, 13–24.

Smits, J.R.M., Melssen, W.J., Buydens, L.M.C. and Kateman, G. (1994). Using artificial neural networks for solving chemical problems. Part I: Multi-layer feed-forward networks, *Chemom. Intell. Lab. Systems*, 22, 165–189.

Sridhar, D.V., Seagrave, R.C. and Bartlett, E.B. (1996). Process modeling using stacked neural networks, *AIChE J.*, 42, 2529–2539.

Tatliera, M., Cigizoglub, H.K. and Erdem-Senatalara, A. (2005). Artificial neural network methods for the estimation of zeolite molar compositions that form from different reaction mixtures, *Comput. Chem. Eng.*, 30, 137–146.

Tickle, A.B., Andrews, R., Golea, M. and Diederich, J. (1998). The truth will come to light: Directions and challenges in extracting rules from trained artificial neural networks, *IEEE Trans. Neural Networks*, 9, 1057–1068.

Tompos, A., Margitfalvi, J.L., Tfirst, E. and Végvári, L. (2003). Information mining using artificial neural networks and "holographic research strategy", *Appl. Catal., A: General*, 254, 161–168.

Tompos, A., Margitfalvi, J.L., Tfirst, E., Végvári, L., Jaloull, M.A., Khalfalla, H.A. and Elgarni, M.M. (2005). Development of catalyst libraries for total oxidation of methane: A case study for combined application of "holographic research strategy and artificial neural networks" in catalyst library design, *Appl. Catal., A: General.*, 285, 65–78.

Tompos, A., Margitfalvi, J.L., Tfirst, E. and Végvári, L. (2006). Evaluation of catalyst library optimization algorithms: Comparison of the holographic research strategy and the genetic algorithm in virtual catalytic experiments, *Appl. Catal., A: General*, 303, 72–80.

Umegaki, T., Watanabe, Y., Nukui, N., Omata, K. and Yamada, M. (2003). Optimization of catalyst for methanol synthesis by a combinatorial approach using a parallel activity test and genetic algorithm assisted by a neural network, *Energy Fuels*, 17, 850–856.

Urschey, J., Kühnle, A., and Maier, W.F. (2003). Combinatorial and conventional development of novel dehydrogenation catalysts, *Appl. Catal., A: General*, 252, 91–106.

Valero, S., Argente, E., Botti, V., Serra, J.M., Serna, P., Moliner, M. and Corma. A. (2009). DoE framework for catalyst development based on soft computing techniques, *Comput. Chem. Eng.*, 33, 225–238.

Watanabe, Y., Umegaki, T., Hashimoto, M., Omata, K., Yamada, M. (2004). Optimization of Cu oxide catalysts for methanol synthesis by combinatorial tools using 96 well microplates, artificial neural network and genetic algorithm, *Catal. Today*, 89, 455–464.

White, H. (1992). *Artificial Neural Networks: Approximation and Learning Theory*, Blackwell, Cambridge, 320 p.

Williamson, J.R. (1996). Gaussian ARTMAP: A neural network for fast incremental learning of noisy multidimensional maps, *Neural Networks*, 9, 881–897.

Zahedi, G., Jahanmiri, A. and Rahimpor, M.H. (2005). A Neural Network Approach for Prediction of the CuO-ZnO-Al_2O_3 catalyst deactivation, *Int. J. Chem. Reactor Eng.*, 3, A8.

Zupan, J. and Gasteiger, J. (1993). *Neural Networks for Chemists,* Wiley–VCH, Weinheim, 336 p.

Zupan, J. and Gasteiger, J. (1999). *Neural Networks in Chemistry and Drug Design: An Introduction,* Wiley-VCH, Weinheim, 380 p.

Chapter 7

Tuning Evolutionary Algorithms with Artificial Neural Networks

7.1. Heuristic Parameters of Genetic Algorithms

In Chapter 3, an important property of genetic algorithms was mentioned: due to incorporation of randomness, the optimisation paths can leave the attraction area of the nearest local optimum and continue searching for a global one. The probability that at least one optimisation path reaches the global optimum increases with the diversity of the locations that form the population of the algorithm. On the other hand, all paths depend heavily on the *probability distributions* of the random variables involved (e.g., of the occurrence of mutation and crossover, or of the intensity of quantitative mutation). Their distributions influence all important aspects of the global optimisation: its accuracy (i.e., how close the method gets to the global optimum), convergence speed (how many generations it needs to get there), as well as the diversity of the resulting population. Given the biological inspiration of the random variables, a particular distribution cannot be justified mathematically, but choosing it is a heuristic task. The most crucial heuristic choices entailed by a genetic algorithm are:

(i) *Overall probability of any modification* (crossover, qualitative or quantitative mutation) of an individual.

(ii) *Ratio between the conditional probabilities* of crossover and qualitative or quantitative mutation, conditioned on the occurrence of any modification.

(iii) *Distribution of the intensity of quantitative mutation*, e.g., distribution of the coefficient with which coordinates of locations have to be multiplied/divided.

For example, the first genetic algorithm that was developed specifically for the optimisation of solid catalysts relied on the following heuristic parameters (Wolf *et al.*, 2000):

- A component-type specific probability of any modification through crossover and mutation.
- The ratio between the probabilities of qualitative mutation and crossover.
- The asymptotic ratio between the probability of quantitative mutation and the component-type specific probability of the occurrence of any modification.
- The coefficient of quantitative mutation, used to multiply the quantitatively mutated values.

The population size can also be a matter of heuristic choice. However, in catalytic experiments, population size is usually determined by the number of channels in the reactor in which the catalysts are tested.

The important role of such heuristic parameters in the genetic optimisation of catalytic materials has already been sufficiently recognised (Umegaki *et al.*, 2003; Rodemerck *et al.*, 2004; Serra and Corma, 2004; Clerc *et al.*, 2005; Ohrenberg *et al.*, 2005; Pereira *et al.*, 2005; Tompos *et al.*, 2005; Clerc, 2006; Holeňa, 2006; Rothenberg *et al.*, 2006; Valero *et al.*, 2009). However, if the values of the objective function have to be obtained in a costly experimental way, then emploing that costly evaluation also for tuning the involved heuristic parameters is not affordable. In addition, experimental evaluation of one generation of individuals typically requires timespans in the order of hours to days, even if one excludes the time required to reach a steady state of the catalytic material.

7.2. Parameter Tuning Based on Virtual Experiments

The aim of this chapter is to show that the time and costs needed for the evaluation of empirical objective functions can be substantially reduced by means of an approach that is in optimisation referred to as *surrogate modelling* (Ratle, 2001; Ulmer *et al.*, 2003; Ong *et al.*, 2005; Zhou *et al.*,

2007), or sometimes also *metamodelling* (Emmerich *et al.*, 2002). In the context of the optimisation of catalytic materials, surrogate modelling can be viewed as replacing real experiments with simulated, virtual experiments in a computer. In Baumes *et al.* (2004) and Farrusseng *et al.* (2007), such virtual experiments are called *virtual screening*; in Valero *et al.* (2009), they are called *pre-screening*.

Although surrogate modelling is a general optimisation approach, it is most frequently encountered in connection with evolutionary optimisation — thus in connection with exactly the kind of optimisation that is relevant to catalysis (cf. Chapters 3 and 4). The reason for this is that in evolutionary optimisation, the approach leads to the approximation of the landscape of the fitness function, i.e., to a method that is known to be useful in general (Ratle, 1998; Jin, 2005). Principally, surrogate modelling consists in restricting empirical evaluations to locations that are expected to be close to the global optimum according to some regression model obtained from the available data, which is called a *surrogate model* of the empirical objective function. Accordingly, surrogate modelling has the following main steps:

(i) Collecting an initial set of locations in which the objective function has already been empirically evaluated. This can be the first generation or several first generations of the evolutionary algorithm, but such locations are frequently available in advance.
(ii) Approximating the objective function by a surrogate model, with the use of the set of all locations in which the function has already been empirically evaluated.
(iii) Running the evolutionary method for a population considerably larger than the desired population size, with the empirical objective function replaced by the surrogate model.
(iv) Forming the next generation of the desired size as a subset of the large population obtained in the preceding step. The next generation includes on the one hand, locations which are estimated, according to the surrogate model used, to best indicate the global optimum, and on the other hand, locations that most contribute to the diversity of the population, to allow the evolutionary method to continue the search for the global optimum in diverse parts of the input space.

(v) Empirically evaluating the objective function in all locations that belong to the next generation of the desired size, and returning to step (ii).

Alternatively to steps (ii)–(v), the algorithm can be run for the desired population size only, interleaving each generation in which the original objective function is empirically evaluated with a certain number of generations in which the surrogate model is evaluated.

The cost of evaluating a surrogate model in a population of locations, even for a larger number of combinations of values of heuristic parameters, is negligible compared with the costs of experimentally evaluating the original objective function in those locations for any single combination of their values.

What does this approach mean in the context of genetic optimisation of catalytic materials? If there is no need to provide experimental results for the catalysts proposed in the course of the GA optimisation, then we can proceed with that optimisation for as many generations as desired, and also the number of simultaneous optimisation paths can be chosen as large as desired. Even more importantly, the genetic algorithm can be run several times for *different population sizes,* and its *convergence to the global maximum* of the objective function, as well as the decrease in *diversity of catalyst compositions* can be observed. In the same way, the influence of *other heuristic parameters* of the genetic algorithm on the convergence speed and diversity decrease can also be studied.

One possible choice for a surrogate model involves mappings computed by multilayer perceptrons. Recall from the previous section that these are complicated nonlinear regression models, and a key ingredient of the universal approximation capability of MLPs. Needless to say, training the neural network requires some *initial amount of data* from experimentally evaluating the objective function, to be gathered first. To this end, data from several early generations of the genetic algorithm are usually sufficient, especially if the population size is large. Actually, data from the early generations are more uniformly distributed, making it more likely that the neural network will correctly approximate all optima, including the global optimum. Sometimes, data from other experiments concerning the same problem are also available. Once the parameters have been tuned, they can be used for all the remaining generations of the algorithm, or the tuning can be repeated every several generations, using a new neural network. In the latter case, the data from experimentally evaluating the objective function that were colleted since the last tuning are added to the training data of the network.

If the surrogate model is a mapping computed by an MLP, then according to the preceding chapter, that model is precisely described mathematically. Theoretically, this entails that also its maxima can be localised with an arbitrary precision, differently from the localisation of maxima of the empirical objective function, which is always subject to measurement errors. Moreover, if in the model any of the following activation functions hyperbolic tangens, arctangens or logistic activation function is used, which were recalled in the preceding chapter, then the model has partial derivatives of all orders. Consequently, the maxima can be searched using optimisation methods that require partial derivatives up to the second order. Recall from Chapter 3 that methods of this kind are the fastest and most accurate in seeking the maxima, but have the disadvantage of generally finding only the local maximum in the attraction area of which the optimization path starts. To increase the likelihood that such an optimisation method will find the global maximum of the objective function, the method can be run repeatedly from random starting locations. The highest among all maxima found is then considered as the global one.

7.3. Case Study with the Oxidative Dehydrogenation of Propane

In this section, the above approach is applied to the data concerning 328 materials tested as catalysts for the oxidative dehydrogenation of propane to propene (ODP). The materials consisted of an alpha-aluminium support and mostly a shell of metal-oxide mixtures selected from the oxides of B, Fe, Ga, La, Mg, Mn, Mo, V; partly the catalytic material was also in the porous structure. All of them were tested at the same standard conditions, T = 773K, $p_{C_3H_8}$ = 30,3kPa, p_{O_2} = 10,1kPa and p_{N_2} = 60,6kPa. Details concerning the properties of those materials and their selection and testing have been described in Buyevskaya *et al.* (2001).

When seeking an appropriate surrogate model, multilayer perceptrons with one hidden layer containing 1–20 neurons were considered. Each such MLP had seven input neurons, corresponding to variables that record the molar proportions of seven out of the above-mentioned eight metal oxides in the active shell of the catalytic material (the proportion of the remaining one being implied by the fact that all proportions have to sum up to 100%). On the other hand, each MLP had only one output neuron, corresponding to a variable that records the yield of the ODP product: propene. The generalisation ability of those neural networks on the available ODP data was investigated and it was found that the best average generalisation was achieved by MLPs with the architecture (7,5,1). Therefore, an MLP with that architecture was subsequently trained with all ODP data, using the Levenberg-Marquardt method in combination with early stopping (cf. the preceding chapter). The mapping computed by that MLP was then employed as the surrogate model in this case study.

First of all, the generalisation ability of the model was validated for ten additional catalytic materials, which were not available among the 328 original ones. The results are shown in Table 7.1.

The surrogate model was then used to tune the heuristic parameters of the first genetic algorithm developed specifically for the optimisation of solid catalysts (cf. Section 7.1). To each of those parameters, two to four values have been assigned, according to Table 7.2, resulting in 96 different value combinations altogether. For each value combination, the following results were recorded:

- The convergence of the algorithm during the first 50 generations, measured as the evolution of the highest propene yield among the catalytic materials proposed up to the current generation.
- The decrease in diversity among the retained catalytic materials during the first 50 generations, where diversity is measured as the difference between the highest propene yield and the mean propene yield of all the retained materials.

Table 7.1. Comparison of the predicted and measured propene yield values on additional catalytic materials for the surrogate model.

Proportion in the active shell of catalyst (%)								Propene yield (%)	
B	Fe	Ga	La	Mg	Mn	Mo	V	Predicted	Measured
0	0	27	0	0	0	26	47	5.5	6.2
0	0	19	0	72	0	9	0	0.02	1.4
0	62	8	0	0	0	14	16	5.6	6.4
0	20	0	0	48	0	16	16	5.4	5.6
0	22	0	24	0	23	31	0	3.0	2.2
0	0	84	0	8	0	0	8	6.0	7.5
0	13	11	0	28	0	10	38	6.1	7.3
0	33	34	0	18	0	15	0	2.7	2.5
15	7	28	0	0	50	0	0	4.0	3.8
0	0	24	0	0	27	7	43	7.3	7.4

Table 7.2. Values of the heuristic parameters of the genetic algorithm considered when investigating the influence of those properties on the convergence speed and diversity decrease of the algorithm.

Heuristic parameter	Considered values
Population size	28, 56, 112, 280
Probability of any modification	10%, 90%
Probability ratio "qualitative mutation/crossover"	0.1, 1, 10
Asymptotic probability ratio "quantitative mutation/any modification"	0.05, 0.75
Coefficient of quantitative mutation	2, 10

When investigating the convergence of the algorithm, attention was paid not only to its speed, but also to comparing the highest propene yield among the catalytic materials proposed up to the current generation with the global maximum of the propene yield assumed by the surrogate model. To this end, use was made of the facts recalled above: that the global maximum of the surrogate model can be found (at least in theory), and that it can be found by means of optimisation methods using partial derivatives up to the second order. Among them, the Levenberg-Marquardt method was chosen, which was recalled in Chapter 3. Because this method searches only for local maxima in the seven-dimensional space in which the surrogate model was constructed, it was started from seven different starting locations. Two of these correspond to catalytic materials with compositions 40%Ga34%Mg11%Mo15%V and 32%Ga17%Mg19%Mn4%Mo28%V; these were the two best among the available catalysts, i.e., the ones for which the highest propene yield was achieved. The other five locations correspond to pure oxides of Ga, Mg, Mn, Mo and V. For four of those seven starting locations, including the locations that correspond to the two best catalysts, the optimisation method found the same local maximum 9.09% of propene yield in the same location corresponding to the compositions 21%Ga40%Mg21%Mn18%V. For the remaining three starting locations, three different local maxima 8.9%, 7.5% and 5.2% of propene yield were found. Therefore, the local maximum 9.09% in the above mentioned point is being considered in the sequel as the global maximum of the propene yield assumed by the surrogate model.

Consequently, the genetic algorithm needs to be run for 50 generations with any of the considered combinations of values of the adjustable parameters. This takes a considerable amount of time, especially for larger population sizes. For example, running the algorithm for 50 generations in the case of the largest population size of 280 compositions took approximately three hours on a personal computer with a 2 GHz CPU frequency, using an Oracle database for storing the results. Therefore, we have run the algorithm only once for each combination.

Under these circumstances, the results obtained do not provide any information about the extent of random influences present in the algorithm, most substantially in the choice of catalysts in the first generation. To reduce that shortcoming, two measures were taken:

1. All runs with an identical population size shared their first generation. Hence, at least the random influences pertaining to the first generation did not interfere with the influence of heuristic parameters of the algorithm, with the exception of its population size.
2. To gain at least some idea about the variance due to random influences, one particular combination of values of the adjustable parameters, with a population size of 28, was run ten times. The variability of those ten runs is summarised in Figure 7.1 (convergence

of the algorithm) and Figure 7.2 (decrease of population diversity). It can be seen that:

- The variability starts to be approximately constant already in early generations.
- The mean absolute error and standard deviation of the maximal propene yield represent relative errors of approximately 0.7% and 0.8% with respect to the global maximum.

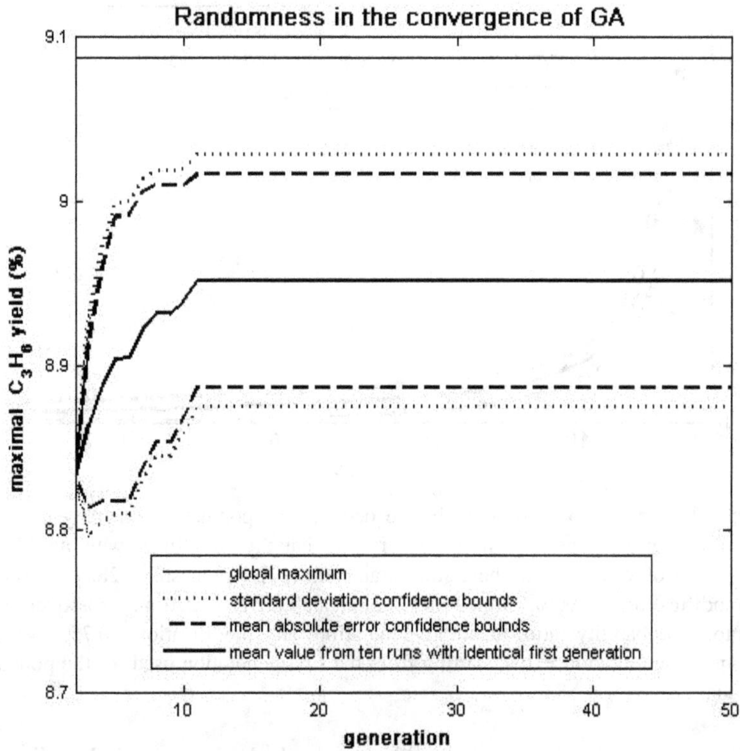

Figure 7.1. Example of variablity in convergence of the genetic algorithm caused by random influences. The algorithm was run repeatedly ten times with an identical combination of values of the adjustable parameters (population size = 28, probability of any modification = 90%, probability ratio qualitative mutation: crossover = 10, asymptotic probability ratio quantitative mutation: any modification = 0.75, coefficient of quantitative mutation = 10), starting from the first generation used for the population size equal 28.

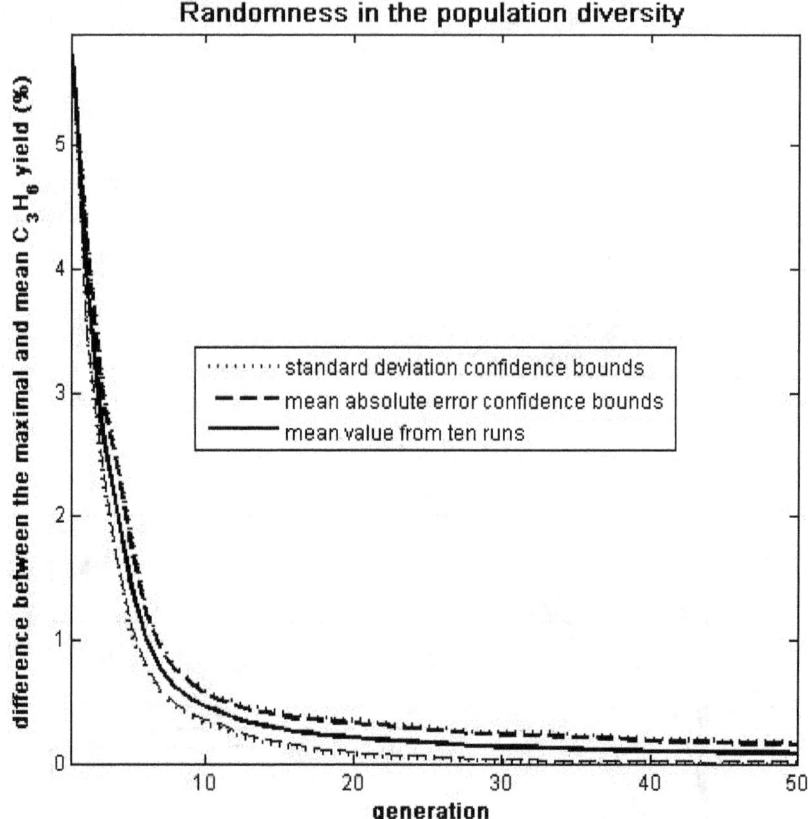

Figure 7.2. Example of variability in the decrease of population diversity caused by random influences. The algorithm was run repeatedly ten times with an identical combination of values of the heuristic parameters (population size = 28, probability of any modification = 90%, probability ratio qualitative mutation: crossover = 10, asymptotic probability ratio quantitative mutation: any modification = 0.75, coefficient of quantitative mutation = 10), starting from the first generation used for the population size equal 28.

- The mean absolute error and standard deviation of the diversity among the retained catalysts, i.e., the mean absolute error and standard deviation of the difference between the highest propene yield and the mean propene yield of all the retained catalysts, are 0.07% and 0.09%, respectively. It is worth noticing that these values are smaller than the typical experimental accuracy of catalytic mesurements, which is 0.1%.

Table 7.3. Generation in which the highest propene yield surpassed values selected as indicators of convergence (or "—" if the highest yield did not surpass them up to the 50th generation). Results for all combinations of the considered values of heuristic parameters.

Probability of any modification	Probability ratio qualitative mutation: crossover	Asymptotic probability ratio quantitative mutation: any modification	Coefficient of quantitative mutation	Population size											
				28			56			112			280		
				Values selected as indicators of convergence											
				Middle value of global max. and max. of first generation	Global max. — relative error 0.8%	9.05%	Middle value of global max. and max. of first generation	Global max. — relative error 0.8%	9.05%	Middle value of global max. and max. of first generation	Global max. — relative error 0.8%	9.05%	Middle value of global max. and max. of first generation	Global max. — relative error 0.8%	9.05%
10%	0.1	0.05	2	5	9	—	5	—	—	8	9	—	4	46	—
			10	—	—	—	3	—	—	6	49	49	4	9	—
		0.75	2	5	5	—	5	—	—	5	5	5	3	7	—
			10	—	—	—	6	—	—	—	—	—	3	6	—
	1	0.05	2	7	—	—	5	13	13	8	—	—	4	20	—
			10	3	3	3	4	—	—	5	—	—	3	10	—
		0.75	2	—	—	—	4	—	—	4	8	8	3	7	—
			10	—	—	—	—	—	—	4	4	—	4	—	—
	10	0.05	2	—	—	—	8	—	—	5	—	—	3	5	9
			10	6	—	—	10	—	—	6	12	—	3	8	—
		0.75	2	8	—	—	3	28	—	2	2	2	3	4	—
			10	4	4	—	5	—	—	8	—	—	3	7	7
90%	0.1	0.05	2	5	5	—	3	—	—	4	—	—	5	46	46
			10	6	—	—	7	—	—	7	22	—	4	39	—
		0.75	2	—	—	—	5	—	—	5	5	5	3	5	6
			10	6	6	—	5	—	—	4	5	9	2	7	—
	1	0.05	2	3	7	—	3	—	—	5	5	—	3	8	—
			10	—	—	—	5	—	—	4	12	—	3	6	6
		0.75	2	5	—	—	4	—	—	5	7	7	4	8	—
			10	—	—	—	4	—	—	5	—	—	3	—	—
	10	0.05	2	6	—	—	3	18	—	6	10	—	3	—	—
			10	8	—	—	5	—	—	3	12	13	3	6	11
		0.75	2	6	6	—	5	—	—	7	8	—	4	—	—
			10	6	6	—	5	—	—	3	9	9	3	—	—

Table 7.4. Generation since which the diversity of catalysts was (up to the 50th generation) below values selected as indicators of decreasing diversity (or "—" if the diversity was above those values in the 50th generation). Results for all combinations of the considered values of heuristic parameters.

Probability of any modification	Probability ratio qualitative mutation: crossover	Asymptotic probability ratio quantitative mutation: any modification	Coefficient of quantitative mutation	Population size: 28			56			112			280		
				Values selected as indicators of decreasing diversity											
				Half diversity of first generation	1.0%	0.1%	Half diversity of first generation	1.0%	0.1%	Half diversity of first generation	1.0%	0.1%	Half diversity of first generation	1.0%	0.1%
10%	0.1	0.05	2	3	7	—	3	6	13	3	7	19	3	6	13
			10	3	5	9	3	6	24	3	7	13	4	7	15
		0.75	2	3	4	—	3	4	12	3	6	14	3	5	13
			10	3	7	8	3	5	11	3	6	11	3	5	13
	1	0.05	2	4	6	—	3	6	—	3	16	—	3	7	18
			10	3	5	14	3	7	18	3	7	20	3	7	19
		0.75	2	3	5	11	3	4	13	3	6	11	3	5	12
			10	4	6	10	3	5	10	3	7	12	3	5	12
	10	0.05	2	3	7	32	3	6	—	3	8	—	4	9	50
			10	4	8	27	3	7	33	3	10	—	4	9	36
		0.75	2	3	5	16	3	7	41	4	8	16	4	7	15
			10	3	8	22	3	8	21	3	7	14	3	8	21
90%	0.1	0.05	2	3	7	—	3	5	21	3	7	17	3	7	—
			10	3	5	—	3	6	15	3	7	15	3	6	—
		0.75	2	3	4	6	3	4	9	3	6	12	3	6	11
			10	3	7	16	3	5	10	3	6	14	3	6	10
	1	0.05	2	3	6	—	3	5	12	3	8	21	4	6	29
			10	3	5	8	3	6	8	3	7	18	4	7	17
		0.75	2	3	4	11	3	4	9	3	6	14	3	6	12
			10	3	5	15	3	5	11	3	8	15	4	6	9
	10	0.05	2	3	7	—	3	7	—	3	11	—	4	9	37
			10	3	8	—	3	7	42	3	8	—	4	10	46
		0.75	2	3	5	—	3	6	42	3	9	18	3	7	17
			10	3	7	—	3	7	35	3	10	32	4	7	15

Finally, the overall results of running the genetic algorithm for all 96 value combinations of the adjustable parameters are summarised in Table 7.3 (convergence of the algorithm) and Table 7.4 (decrease of catalyst diversity). In Table 7.3, the following indicators of convergence have been used:

(i) The middle value of the highest propene yield among the catalysts proposed in the first generation and the global maximum of propene yield, i.e., the value ½(highest propene yield in first generation + propene yield global maximum). Hence, this indicator reduces the bias due to the fact that all value combinations with identical population size shared their first generation.

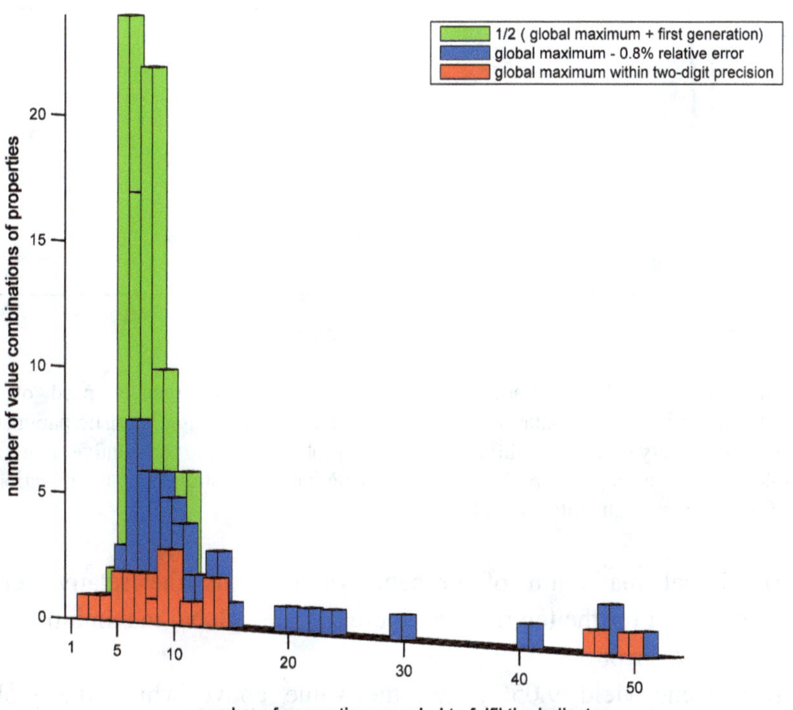

Figure 7.3. Distribution of the number of combinations of the considered values of properties used as heuristic parameters, according to the generations in which the genetic algorithm fulfilled the three considered indicators of convergence.

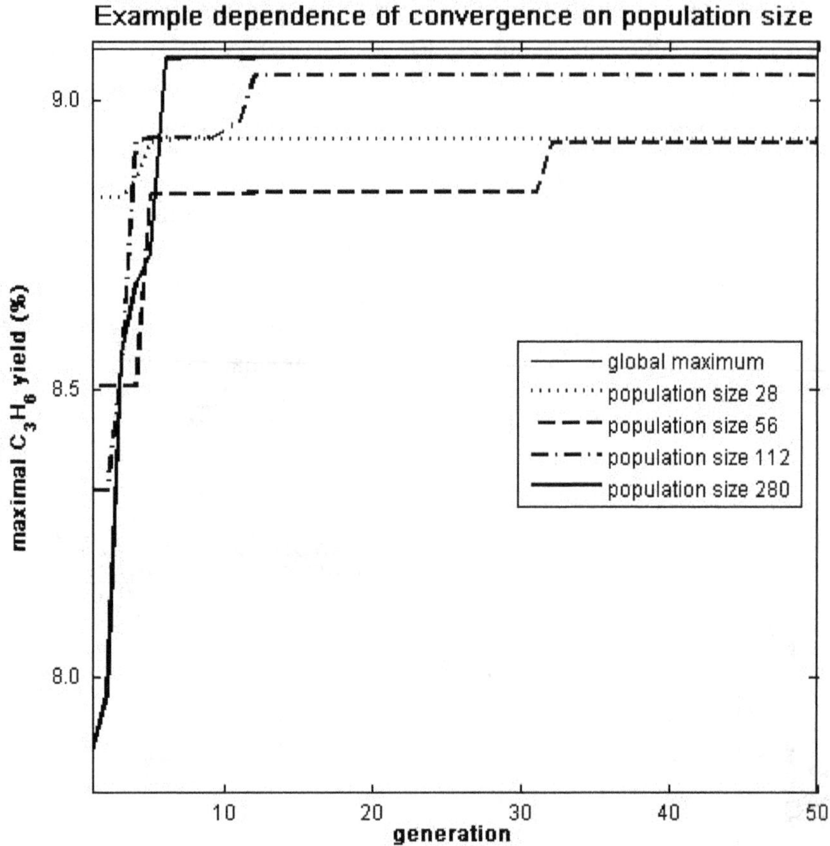

Figure 7.4. Example illustrating the increase of the convergence speed of the genetic algorithm with population size. The values of the remaining heuristic parameters were: probability of any modification = 90%, probability ratio qualitative mutation: crossover = 1, asymptotic probability ratio quantitative mutation: any modification = 0.05, coefficient of quantitative mutation = 10.

(ii) Global maximum of propene yield minus the relative error recorded for the ten runs in Figure 7.1, i.e., global maximum minus relative error 0.8%.

(iii) Propene yield 9.05%, i.e., the value above which the global maximum of propene yield has been reached within two-digit precision.

In Table 7.4, the following indicators of diversity decrease have been used:

(i) Half the diversity of the first generation. Again, this indicator reduces the bias due to the fact that all value combinations with identical population size started from the same first generation.
(ii) Diversity 1%.
(iii) Diversity 0.1%.

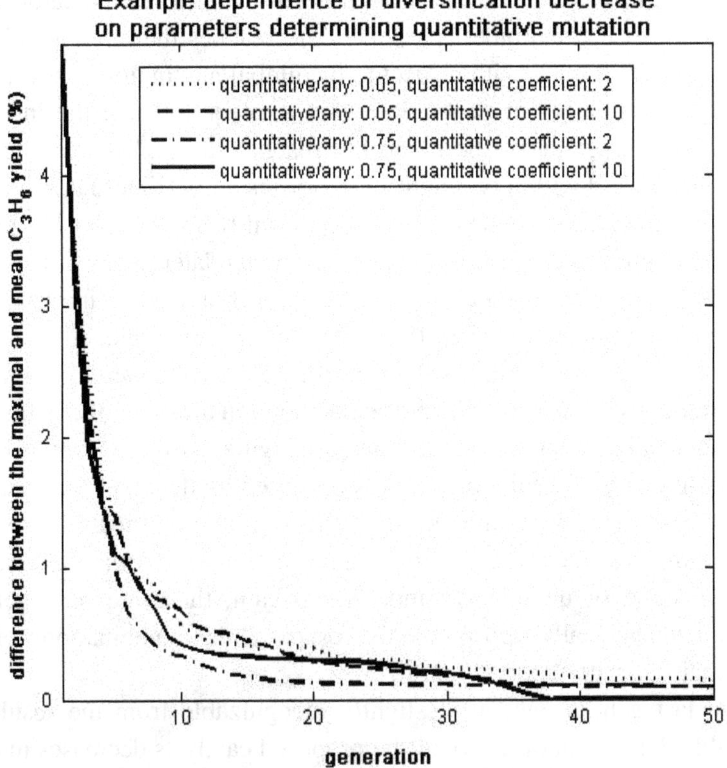

Figure 7.5. Example illustrating that the diversity of the catalysts proposed by the genetic algorithm decreases more with increasing asymptotic probability ratio of quantitative mutation to any modification, as well as with increasing coefficient of quantitative mutation. The values of the remaining heuristic parameters were: population size = 56, probability of any modification = 90%, probability ratio qualitative mutation: crossover = 1.

From the results in Table 7.3 and Table 7.4, the following four conclusions can be drawn:

1. The indicators of convergence that are fulfilled for a given combination of values of adjustable parameters are typically already fulfilled after early generations. More precisely, the mean number of generations that the algorithm needs to fulfil the indicator (i) of convergence (the highest propene yield increases above the middle of its value in the first generation and the global maximum) is four, the mean number of generations that it needs to fulfil the indicator (ii) of convergence (the highest propene yield reaches the range within the relative error 0.8% with respect to the global maximum) is seven, and the mean number of generations that it needs to fulfil the indicator (iii) of convergence (the algorithm reaches the global maximum of propene yield within two-digit precision) is eight (Figure 7.3). Hence, early generations are much more important from the point of view of genetic optimisation of the propene yield than later generations.
2. The convergence speed of the genetic algorithm tends to increase with increasing population size. However, this is only a general tendency, which interferes with the influence of the remaining heurisitc parameters. It is only for particular combinations of them that the tendency becomes really apparent (see Figure 7.4 for an example).
3. The diversity of catalytic materials proposed by the algorithm tends to decrease more with an increase of the asymptotic probability ratio of quantitative mutation to any modification, and with increasing coefficient of quantitative mutation. Again, these are only general tendencies, really apparent only for particular combinations of the remaining parameters (Figure 7.5).
4. The last general tendency, slightly recognizable from the results in Table 7.4, is that diversity of the proposed catalysts decreases more if crossover and qualitative mutation occur with equal probability than if one of those probabilities is substantially higher than the other (Figure 7.6).

Finally, the results in Table 7.3 and Table 7.4 show that all the adjustable parameters strongly interact in their influence; therefore, an improvement in the GA behaviour is possible only if all those parameters

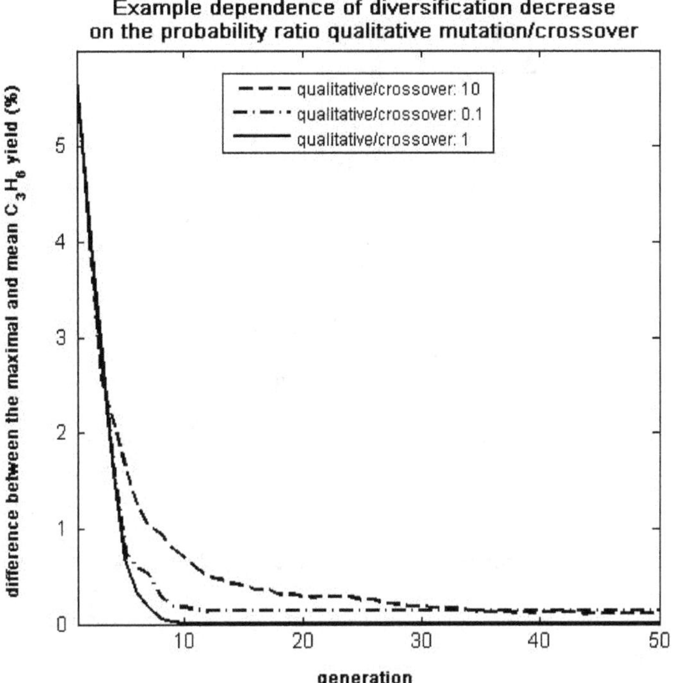

Figure 7.6. Example illustrating that the diversity of the catalytic materials proposed by the genetic algorithm decreases more if crossover and qualitative mutation occur with equal probability than if one of them substantially prevails. The values of the remaining heuristic parameters were: population size = 28, probability of any modification = 90%, asymptotic probability ratio quantitative mutation: any modification = 0.05, coefficient of quantitative mutation = 10.

are tuned simultaneously. In particular, these results suggest, for each considered population size, one or several combinations of values of heuristic parameters of the algorithm that are most appropriate from the point of view of how many generations it takes to the algorithm to find a catalytic material that leads to a propene yield close to the global maximum. Consequently, it is suggested that such a combination of parameter values is used for the considered population size in the subsequent generations of the genetic algorithm within the same experiment. Four suggested combinations (one for each population size) are listed in Table 7.5, with the convergence of the algorithm for them depicted in Figure 7.7. It should be emphasised that the suggested

Table 7.5. Combinations of values of heuristic parameters that the results in Table 7.3 and Table 7.4 suggest to be used for each of the considered population sizes.

Population size	Values of heuristic parameters
28	Probability of any modification = 10%, probability ratio qualitative mutation: crossover = 1, asymptotic probability ratio quantitative mutation: any modification = 0.05, coefficient of quantitative mutation = 10
56	Probability of any modification = 10%, probability ratio qualitative mutation: crossover = 1, asymptotic probability ratio quantitative mutation: any modification = 0.05, coefficient of quantitative mutation = 2
112	Probability of any modification = 10%, probability ratio qualitative mutation: crossover = 10, asymptotic probability ratio quantitative mutation: any modification = 0.75, coefficient of quantitative mutation = 2
280	Probability of any modification = 90%, probability ratio qualitative mutation: crossover = 1, asymptotic probability ratio quantitative mutation: any modification = 0.05, coefficient of quantitative mutation = 10

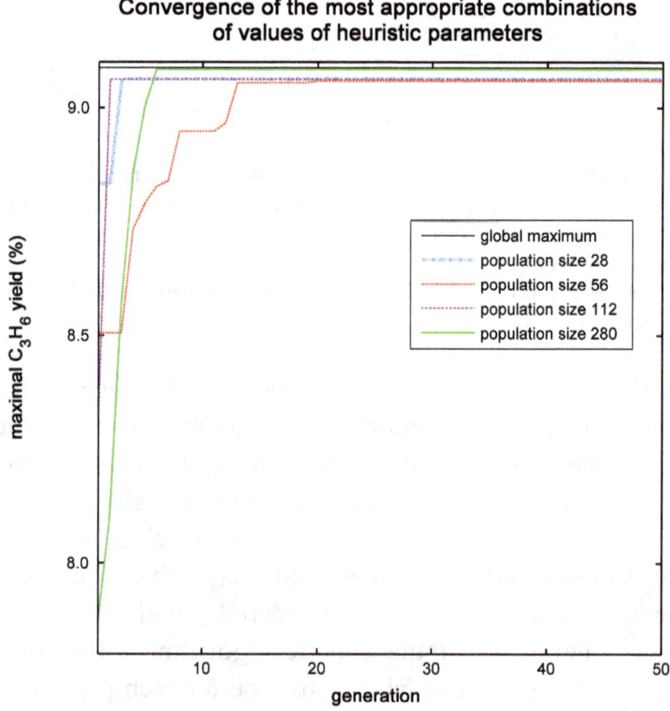

Figure 7.7. Convergence of the genetic algorithm for the combinations of values of heuristic parameters suggested in Table 7.5

combinations can be safely used only for the respective population sizes or for population sizes that are very close to them. Choosing the most appropriate value combinations for other population sizes requires performing a separate investigation for them, based on additional runs of the genetic algorithm.

Bibliography

Baumes, L.A., Farrusseng, D., Lengliz, M. and Mirodatos, C. (2004). Using artificial neural networks to boost high-throughput discovery in heterogeneous catalysis, *QSAR Comb. Sci.*, 23, 767–778.

Buyevskaya, O.V., Bruckner, A., Kondratenko, E.V., Wolf, D. and Baerns, M. (2001). Fundamental and combinatorial approaches in the search for and optimization of catalytic materials for the oxidative dehydrogenation of propane to propene, *Catal. Today,* 67, 369-378.

Clerc, F., Lengliz, M., Farrusseng, D., Mirodatos, C., Pereira, S.R.M. and Rakotomata, R. (2005). Library design using genetic algorithms for catalysts discovery and optimization, *Rev. Sci. Instrum.*, 76, 062208.

Clerc, F. (2006). Optimization and data mining for catalysts library design, Thesis, Université Lyon 1, 189 p.

Emmerich, M., Giotis, A, Özdenir, M., Bäck, T. and Giannakoglou, K. (2002). Metamodel-assisted evolution strategies. In: Guervos, J.J.M. *et al.* (eds.), *Parallel Problem Solving from Nature*, Springer, Berlin, pp. 371–380.

Farrusseng, D., Clerc, F., Mirodatos, C., Azam, N., Gilardoni, F., Thybaut, J.W., Balasubramaniam, P. and Marin, G.B. (2007). Development of an integrated informatics toolbox: HT kinetic and virtual screening, *Comb. Chem. High Throughput Screening*, 10, 85–97.

Jin, Y. (2005). A comprehensive survey of fitness approximation in evolutionary computation, *Soft Comput.* 9, 3–12.

Ohrenberg, A., Törne, C., Schuppert, A. and Knab, B. (2005). Application of data mining and evolutionary optimization in catalyst discovery and high-throughput experimentation: Techniques, strategies, and software, *QSAR Comb. Sci.*, 24, 29–37.

Ong, Y.S., Nair, P.B., Keane, A.J. and Wong, K.W. (2005). Surrogate-assisted evolutionary optimization frameworks for high-fidelity engineering design problems. In Jin, Y. (ed.), *Knowledge Incorporation in Evolutionary Computation*, Springer: Berlin, pp. 307–331.

Pereira, R.M.; Clerc, F.; Farrusseng, D.; Waal, J.C.; Maschmeyer, T. (2005). Effect of the genetic algorithm parameters on the optimisation of heterogeneous catalysts, *QSAR Comb. Sci.*, 24, 45–57.

Ratle, A. (1998). Accelerating the convergence of evolutionary algorithms by fitness landscape approximation. In: Eiben, A.E. *et al.* (eds.), *Parallel Problem Solving from Nature*, Springer: Berlin, pp. 87–96.

Ratle, A. (2001). Kriging as a surrogate fitness landscape in evolutionary optimization, *Artif. Intell. Eng. Des. Anal. Manuf.*, 15, 37–49.

Rodemerck, U., Baerns, M., Holeňa, M. and Wolf, D. (2004). Application of a genetic algorithm and a neural network for the discovery and optimization of new solid catalytic materials, *Appl. Surf. Sci.*, 223, 168–174.

Rothenberg, G., Hageman, J.A., Clerc, F., Frühauf, H.W. and Westerhuis, J.A. (2006). How to find the best homogeneous catalyst? *Catal. Org. React.*, 115, 261–270.

Serra, J.M. and Corma, A. (2004). Two exemplified combinatorial approaches for catalytic liquid-solid and gas-solid processes in oil refining and fine chemicals, In: Hagemeyer, A., Strasser, P. and Volpe, A.F. (eds.), *High-Throughput Screening in Chemical Catalysis*, Wiley-WCH, Weinheim, p. 129–151.

Tompos, A., Margitfalvi, J.L., Tfirst, E., Végvári, L., Jaloull, M.A., Khalfalla, H.A. and Elgarni, M.M. (2005). Development of catalyst libraries for total oxidation of methane: A case study for combined application of "holographic research strategy and artificial neural networks" in catalyst library design, *Appl. Catal., A: General.*, 285, 65–78.

Ulmer, H., Streichert, F. and Zell, A. (2003). Model-assisted steady state evolution strategies. In: Cantú-Paz, E. *et al.* (eds.), *Genetic and Evolutionary Computation — GECCO 2003*, Springer: Berlin, pp. 610–621.

Umegaki, T., Watanabe, Y., Nukui, N., Omata, K. and Yamada, M. (2003). Optimization of catalyst for methanol synthesis by a combinatorial approach using a parallel activity test and genetic algorithm assisted by a neural network, *Energy Fuels*, 17, 850–856.

Urschey, J., Kühnle, A., and Maier, W.F. (2003). Combinatorial and conventional development of novel dehydrogenation catalysts, *Appl. Catal., A: General*, 252, 91–106.

Valero, S., Argente, E., Botti, V., Serra, J.M., Serna, P., Moliner, M. and Corma. A. (2009). DoE framework for catalyst development based on soft computing techniques, *Comput. Chem. Eng.*, 33, 225–238.

Wolf, D., Buyevskaya, O.V. and Baerns, M. (2000). An evolutionary approach in the combinatorial selection and optimization of catalytic materials, *Appl. Catal., A: General*, 200, 63–77.

Zhou, Z.Z., Ong, Y.S., Nair, P.B., Keane, A.J. and Lum, K.Y. (2007). Combining global and local surrogate models to accellerate evolutionary optimization, *IEEE Trans. Syst. Man Cybern. Part C Appl. Rev.*, 37, 66–76.

Chapter 8

Improving Neural Network Approximations

8.1. Importance of Choosing the Right Network Architecture

To ensure the quality of approximation of the unknown dependence $\boldsymbol{D}$ by the function $\boldsymbol{F}$ computed by an artificial neural network, the choice of the number of hidden neurons is crucially important. This was explained in the discussion of the approximation capability of neural networks in Chapter 6, together with the fact that it is sufficient to choose the number of hidden neurons within one hidden layer. On the other hand, this does not prevent using a higher number of hidden layers instead of only one because the set of mappings computed by networks with a particular number of input and output neurons and one hidden layer is a subset of mappings computed by networks with the same number of input and output neurons and more hidden layers.

Therefore, a sufficiently good approximation capability of artificial neural networks cannot be achieved without choosing a network architecture suitable enough with respect to that capability.

A general method for testing the suitability of a particular ANN architecture with respect to the approximation capability is *cross-validation*, more precisely ***k****-fold cross-validation* ($\boldsymbol{k} \geq 2$). This can be equally well used to test the suitability of any particular choice of any ANN property that has to be selected from various possibilities, e.g., a particular choice of the activation function. Cross-validation is actually a general method for choosing parameters and other properties of statistical models (Hand, 1997; Berthold and Hand, 2002). In the context of ANNs trained with catalytic data, the method proceeds as follows:

The set of available data about catalytic materials is randomly partitioned into $\boldsymbol{k}$ parts of approximately equal size.
With the considered architecture, $\boldsymbol{k}$ neural nets are trained, using for each of them one specific part of the obtained partition as test data and all the remaining $\boldsymbol{k}$-1 parts as training data.
To assess the appropriateness of the considered architecture with respect to the approximation of the unknown dependence, the MSE values for the test data are averaged over those $\boldsymbol{k}$ trained neural nets.

Observe that the proportion of catalysts used for training in this method is $1-1/\boldsymbol{k}$. Hence, the higher is $\boldsymbol{k}$, the more information about the available data has been used for training the neural network, and is then inherently incorporated in it. Consequently, the networks using the largest amount of information result from a $\boldsymbol{k}$-fold cross-validation with $\boldsymbol{k}$ equal to the size of available data. Then data about only one single object (e.g., one single catalytic material) are left out for testing. For that reason, this extreme case of cross-validation is also called *leave-one-out* validation.

8.2. Influence of the Distribution of Training Data

In this section, we wish to explain that the lack of data in some areas of the input space can cause a deterioration in the results of ANN training. Recall from Chapter 6 that a neural network is trained with a sequence of training pairs $(\boldsymbol{x}_1,\boldsymbol{y}_1),(\boldsymbol{x}_2,\boldsymbol{y}_2),\ldots,(\boldsymbol{x}_p,\boldsymbol{y}_p)$. In each pair $(\boldsymbol{x}_j,\boldsymbol{y}_j)$, $\boldsymbol{x}_j$ is a vector of values of the input variables characterising catalytic materials, such as proportions of individual components, whereas $\boldsymbol{y}_j$ is the value of some performance measure of the mateiral, e.g., yield or conversion. Alternatively, $\boldsymbol{y}_j$ can be a vector of values of several such measures. Recall also that the error functions reviewed in Chapter 6 decompose into contributions of individual training pairs $(\boldsymbol{x}_j,\boldsymbol{y}_j)$. However, the vectors $\boldsymbol{x}_1,\ldots, \boldsymbol{x}_p$ are not distributed uniformly in the input space of the network: in some areas there are many of them, whereas they are rare in other large parts of the space. If, in an area with a dense occurrence of training vectors, the unknown dependence $\boldsymbol{D}$ is continuous and does not change too fast, the contributions to the overall error of all pairs $(\boldsymbol{x}_j,\boldsymbol{y}_j)$

with $\boldsymbol{x}_j$ from that area are similar. Consequently, the optimisation method employed will not succeed in substantially decreasing the overall error unless it succeeds in decreasing the contributions of training pairs from such areas. That is why after training, the error contributions of training vectors from densely populated areas are typically very low, whereas pairs from sparsely populated areas sometimes contribute with quite high errors. Theoretically, this should not cause any serious problem, assuming the training data are a realisation of a random sample (i.e., of a sequence of independent identically distributed random vectors). That assumption, on which the theory of neural networks and their learning substantially relies (White, 1992), has two important consequences for the error measure MSE:

(i) Irrespectively of the size of the training set, the overall MSE is an unbiased estimate of the average squared error of a single training pair. The term "unbiased estimate" means that the average of this estimate coincides with the estimated average squared error of a single pair — formally

$$\int_{R^{n_I}\times R^{n_O}} \frac{1}{p}\sum_{j=1}^{p}\left(F\left(X_j\right)-Y_j\right)^2 d\mu = \int_{R^{n_I}\times R^{n_O}} \left(F\left(X_i\right)-Y_i\right)^2 d\mu . \quad (10)$$

Here, $\boldsymbol{\mu}$ denotes the distribution of the random vector $(\boldsymbol{X}_i,\boldsymbol{Y}_i)$ the realisation of which is $(\boldsymbol{x}_i,\boldsymbol{y}_i)$ (due to the above assumption, μ does not depend on the index of the random vector).

(ii) The minimum of the overall MSE is a strongly consistent estimate of the minimal average squared error of a single training pair, provided both minima are taken over the set of all functions computed by all MLPs with one hidden layer and prescribed numbers $\boldsymbol{n}_I$ of input neurons and $\boldsymbol{n}_O$ of output neurons. The term "strongly consistent estimate" means that with probability 1, the minimum of the overall MSE for p increasing to infinity approaches the estimated minimal average squared error of a single training pair. Formally,

$$\mu\left(\min_F \frac{1}{p}\sum_{j=1}^{p}\left(F\left(X_j\right)-Y_j\right)^2 \xrightarrow{p\to\infty} \min_F \int_{R^{n_I}\times R^{n_O}} \left(F\left(X_i\right)-Y_i\right)^2 d\mu\right)=1. \quad (11)$$

Recall that MSE (or another error measure) is exactly the objective function of the minimisation performed during neural network learning. Hence, if the above assumption that the training data are a realisation of a random sample is justified, then unequal contributions of individual training pairs should not bias the result of learning, provided the training data is large enough. And even if such data is not large enough, then the bias should in average cancel out.

Unfortunately, data from catalytic experiments belong to those data for which the above assumption is rarely justified. Typically, whole groups of different catalytic materials are prepared and tested using the same devices and apparatus. Then there is some correlation between properties of such materials inevitably, and sometimes also between results of their testing, which contradicts the requirement that the data should be realisations of independent random variables. On the other hand, data concerning materials that were prepared and tested using different devices and apparatus typically do not fulfil the condition that the random variables of which they are realisations are identically distributed. That is why the assumption can be far from validity, which is then true also for its consequences (10)–(11). Therefore, a high rate of errors corresponding to training data from sparsely populated areas can substantially bias the results of training an ANN with data from catalytic experiments, and consequently also the approximation capability of the trained network.

8.3. Boosting Neural Networks

Problems due to the high error contribution of some training data are not specific to artificial neural networks. Moreover, they are not even specific to regression — high error rates contributed by particular training data also typically impair the results of classification. It was in the area of classification that a method called *boosting* was developed to tackle such problems (Schapire, 1990). Its principle consists in developing the classifier iteratively, and increasing the relative influence of the training data that most contributed to errors in the previous iterations on its development in the subsequent iterations. The usefulness of boosting for classification has lead to its extension for regression (Drucker, 1997). Both for classification and for regression, the basic approach to increasing the relative influence of particular training data is resampling the training data according to a distribution that gives them a higher probability of occurrence. This is equivalent to reweighting the contributions of the individual training pairs $(\boldsymbol{x}_j,\boldsymbol{y}_j)$, with higher weights corresponding to higher values of the error measure. The choice between resampling and reweighting is, therefore, typically a matter of the ease of their implementation.

Since ANNs compute regression models, any method for regression boosting is suitable for them (such as the methods proposed in Drucker, 1997; Freund and Schapire, 1997; Friedman *et al.*, 2000; Friedman, 2001; Zemel and Pitassi, 2001; Shrestha, 2006). In the following, the method *AdaBoost.R2* will be explained in detail, as proposed in Drucker (1997). This belongs to adaptive boosting methods, which are the most frequently encountered type of boosting methods (e.g., Schapire, 1990; Drucker, 1997; Altinçay, 2004; Hoffmann, 2004; Redpath and Lebart, 2005; Shrestha, 2006).

In connection with boosting ANNs, it is important to be aware of the difference between the iterations of boosting and the iterations of neural network training. Indeed, boosting iterates on a higher level; one iteration of boosting includes a complete training of an ANN, which can proceed for many hundreds of iterations. Nevertheless, both kinds of iterations are similar in the sense that starting with some iteration, overtraining is present. Therefore, also overtraining introduced through boosting can be reduced with an approach analogous to early stopping: boosting is stopped in that iteration after which the MSE for an independent set of data first increases. Moreover, cross-validation can be used to find the iteration most appropriate for stopping.

Similarly to other adaptive boosting methods, each of the training pairs $(\boldsymbol{x}_1,\boldsymbol{y}_1),\ldots,(\boldsymbol{x}_p,\boldsymbol{y}_p)$ is in the first iteration of AdaBoost.R2 used exactly once. This corresponds to resampling them according to the uniform probability distribution $\boldsymbol{P}_1$ with $\boldsymbol{P}_1(\boldsymbol{x}_k,\boldsymbol{y}_k) = 1/\boldsymbol{p}$ for $\boldsymbol{k} = 1,\ldots,\boldsymbol{p}$. In addition, the weighted average error of the first iteration is set to zero, $\bar{\boldsymbol{E}}_1 = 0$.

In the subsequent iterations $\boldsymbol{i} \geq 2$, the following sequence of steps is performed:

1. A sample $(\boldsymbol{\xi}_1,\boldsymbol{\eta}_1),\ldots,(\boldsymbol{\xi}_p,\boldsymbol{\eta}_p)$ is obtained through resampling $(\boldsymbol{x}_1,\boldsymbol{y}_1),\ldots,(\boldsymbol{x}_p,\boldsymbol{y}_p)$ according to the distribution $\boldsymbol{P}_{i\text{-}1}$.
2. Using $(\boldsymbol{\xi}_1,\boldsymbol{\eta}_1),\ldots,(\boldsymbol{\xi}_p,\boldsymbol{\eta}_p)$ as training data, a regression model $\boldsymbol{F}_i$ is constructed. In our case, $\boldsymbol{F}_i$ is the function from the space of $\boldsymbol{n}_{\mathrm{I}}$-dimensional vectors to the space of $\boldsymbol{n}_{\mathrm{O}}$-dimensional vectors computed by an MLP with given architecture that was trained with $(\boldsymbol{\xi}_1,\boldsymbol{\eta}_1),\ldots,(\boldsymbol{\xi}_p,\boldsymbol{\eta}_p)$.
3. A [0,1]-valued error vector $\boldsymbol{E}_i$ of $\boldsymbol{F}_i$ with respect to $(\boldsymbol{x}_1,\boldsymbol{y}_1),\ldots,(\boldsymbol{x}_p,\boldsymbol{y}_p)$ is calculated as

$$\boldsymbol{E}_i = (\boldsymbol{E}_i(1),\ldots,\boldsymbol{E}_i(\boldsymbol{p})) = \frac{1}{\max\limits_{k=1,\ldots,p}(\boldsymbol{F}_i(\boldsymbol{x}_k)-\boldsymbol{y}_k)^2}\left((\boldsymbol{F}_i(\boldsymbol{x}_1)-\boldsymbol{y}_1)^2,\ldots,(\boldsymbol{F}_i(\boldsymbol{x}_p)-\boldsymbol{y}_p)^2\right).$$

4. The weighted average error of the $\boldsymbol{i}$-th iteration is calculated as

$$\bar{\boldsymbol{E}}_i = \frac{1}{\boldsymbol{p}}\sum_{k=1}^{p}\boldsymbol{P}_i(\boldsymbol{x}_k,\boldsymbol{y}_k)\boldsymbol{E}_i(\boldsymbol{k}).$$

5. Provided $\bar{\boldsymbol{E}}_i < 0.5$, the probability distribution for resampling $(\boldsymbol{x}_1,\boldsymbol{y}_1),\ldots,(\boldsymbol{x}_p,\boldsymbol{y}_p)$ is updated according to

$$\boldsymbol{P}_i(\boldsymbol{x}_k,\boldsymbol{y}_k) = \frac{\boldsymbol{P}_{i-1}(\boldsymbol{x}_k,\boldsymbol{y}_k)\left(\dfrac{\bar{\boldsymbol{E}}_i}{1-\bar{\boldsymbol{E}}_i}\right)^{(1-\boldsymbol{E}_i(k))}}{\sum\limits_{l=1}^{p}\boldsymbol{P}_{i-1}(\boldsymbol{x}_k,\boldsymbol{y}_k)\left(\dfrac{\bar{\boldsymbol{E}}_i}{1-\bar{\boldsymbol{E}}_i}\right)^{(1-\boldsymbol{E}_i(k))}}, \boldsymbol{k} = 1,\ldots,\boldsymbol{p}.$$

   (The case $\bar{\boldsymbol{E}}_i \geq 0.5$ will be discussed below.)
6. As the *boosting approximation* in the $\boldsymbol{i}$-th iteration serves the median of the approximations $\boldsymbol{F}_1,\ldots,\boldsymbol{F}_i$ with respect to the probability distribution proportional to the vector

$$\left(\frac{\bar{E}_1}{1-\bar{E}_1},\ldots,\frac{\bar{E}_i}{1-\bar{E}_i}\right). \tag{12}$$

Needless to say, the boosting approximation can be applied to any input, no matter whether it belongs to training data or to test data. The errors used to assess the quality of the boosting approximation are then called *boosting errors*, e.g., *boosting MSE*, or *boosting MAE* (for simplicity, the approximation in the first iteration, $\boldsymbol{F}_1$, is also called boosting approximation if boosting is performed, and the respective errors are then also called boosting errors, although boosting does not actually introduce any modifications in the first iteration).

The above formulation of the method deals only with the case $\bar{\boldsymbol{E}}_i < 0.5$. For $\bar{\boldsymbol{E}}_i \geq 0.5$, the original formulation of the method in Drucker (1997) proposes to stop the boosting. However, this is not allowed if the stopping criterion should be based on an independent set of validation data because the calculation of $\bar{\boldsymbol{E}}_i$ does not rely on any such independent dataset, but relies solely on the training data. A possible alternative for the case $\bar{\boldsymbol{E}}_i \geq 0.5$ is reinitialisation, i.e., proceeding as in the first iteration (Altinçay, 2004).

8.4. Case Study with HCN Synthesis Continued

Cross-validation and boosting will now be illustrated on the case study with HCN synthesis introduced in Chapter 5. Recall that in this case study, data on a total of 696 catalytic materials were available. It is worth mentioning that in many published applications of artificial neural networks to catalysis, a substantially smaller amount of data was used (Kito *et al.*, 1994; Hou *et al.*, 1997; Huang *et al.*, 2001; Cundari *et al.*, 2001; Corma *et al.*, 2002; Tompos *et al.*, 2003; Tompos *et al.*, 2006; Günay and Yildirim, 2008).

According to Chapter 5, the data gathered in this case study are described by the following variables:

- A discrete input variable recording the support of the catalyst.
- Eleven continuous input variables recording the proportions of the *metal additives* Y, La, Zr, Mo, Re, Ir, Ni, Pt, Zn, Ag, and Au in the active shell of the catalyst, their values are interconnected through the condition that they have to sum up to 100%, that is why only ten of them are really needed.
- Three continuous output variables recording the conversions of CH_4 and NH_3, and the HCN yield.

Each considered neural network then needs to have *14 input neurons* and *three output neurons*. The first four of the input neurons code the material used as support in the manner described in Chapter 6, Subsection 6.2.1. The remaining ten input neurons correspond to the proportions of the 10 metals Y, La, Mo, Re, Ir, Ni, Pt, Zn, Ag, Au (the metal left out was Zr, which is the one least frequent in the available data; notice that the proportion of Zr is nevertheless always derivable due to the fact that the proportions of all active elements sum up to 100%).

For the architecture search by means of cross-validation, only data about catalysts from the first to the sixth generation of the genetic algorithm and about the 52 catalysts with manually designed composition were employed, thus altogether data about 604 catalytic materials. Data about catalysts from the seventh generation were completely excluded and left out for validating the search results. To use as much information as possible from the employed data, cross-validation was applied as the extreme 604-fold variant, i.e., leave-one-out validation. The set of architectures within which the search was performed was delimited by means of the heuristic *pyramidal condition*: the number of neurons in a subsequent layer must not increase the number of neurons in a previous layer. This condition in particular implies:

(i) For MLPs with one hidden layer: $\boldsymbol{n}_{\mathrm{O}} \leq \boldsymbol{n}_{\mathrm{H}} \leq \boldsymbol{n}_{\mathrm{I}}$, in our case $3 \leq \boldsymbol{n}_{\mathrm{H}} \leq 14$ (12 architectures).

(ii) For MLPs with two hidden layers: $\boldsymbol{n}_{\mathrm{O}} \leq \boldsymbol{n}_{\mathrm{H2}} \leq \boldsymbol{n}_{\mathrm{H1}} \leq \boldsymbol{n}_{\mathrm{I}}$, in our case $3 \leq \boldsymbol{n}_{\mathrm{H2}} \leq \boldsymbol{n}_{\mathrm{H1}} \leq 14$ (78 architectures).

The mean values, 10% percentiles and 90% percentiles (10%-iles and 90%-iles, for short) of the cross-validation MSE on test data are summarised for the considered MLP architectures with one hidden layer in Figure 8.1, and for the architectures with two hidden layers in Figure 8.2. For the best five architectures of each kind, ordered according to the average value of the cross-validation MSE on test data, these three statistics are in addition recalled in Table 8.1, complemented by corresponding statistics for the training data.

The results of the architecture search were validated on the data about the 92 catalytic materials from the seventh generation of the genetic algorithm (cf. a similar use of data from a subsequent generation of a genetic algorithm to test networks trained with data from the previous generarions in Corma *et al.* (2002), although without employing cross-validation). For the validation, only one-hidden-layer architectures with the number of hidden neurons n_H from the range $7 \leq n_H \leq 14$, and 2-hidden layers architectures with the numbers of hidden neurons n_{H1} and n_{H2} from the ranges $6 \leq n_{H2} \leq n_{H1} \leq 11$ were considered, due to the fact that the five best architectures of each kind had the numbers of hidden neurons from those ranges (Table 8.1). The validation proceeded as follows:

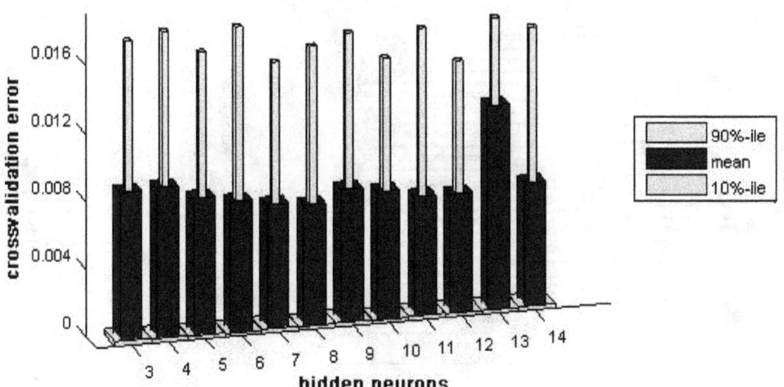

Figure 8.1. Mean values, 10% percentiles and 90% percentiles of the cross-validation MSE on test data for each of the 12 MLP architectures with one hidden layer fulfilling the condition $3 \leq n_H \leq 14$.

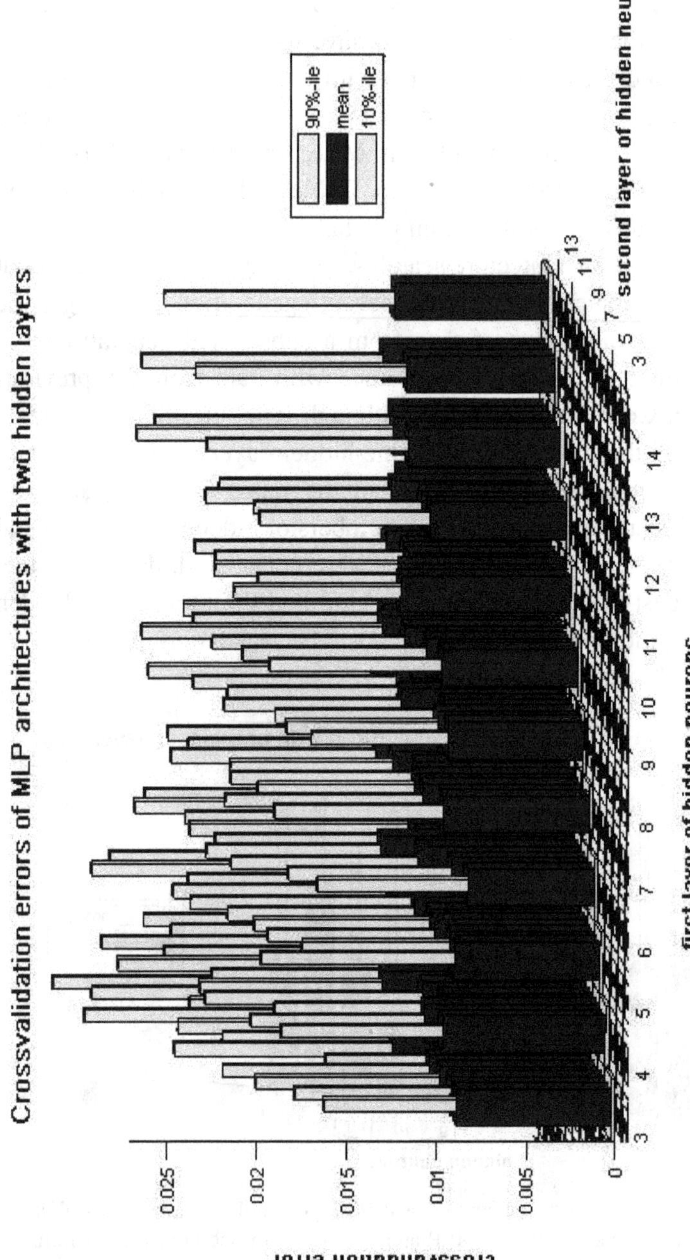

Figure 8.2. Mean values, 10% percentiles and 90% percentiles of the cross-validation MSE on test data for each of the 78 MLP architectures with two hidden layers fulfilling the condition $3 \leq \boldsymbol{n}_{H2} \leq \boldsymbol{n}_{H1} \leq 14$.

Table 8.1. Information about the best five MLP architectures with one hidden and two hidden layers, ordered according to the average value of the cross-validation MSE on test data.

MLP architectures with 1 hidden layer						
Architecture	Cross-validation MSE on the test data [$\cdot 10^{-3}$]			Cross-validation MSE on the training data [$\cdot 10^{-3}$]		
	mean	10%-ile	90%-ile	mean	10%-ile	90%-ile
(14,12,3)	7.09	0.22	14.79	4.60	3.72	5.80
(14,11,3)	7.13	0.26	16.90	4.70	3.79	5.97
(14,8,3)	7.22	0.22	16.40	4.98	4.00	6.50
(14,14,3)	7.38	0.26	16.40	4.44	3.54	5.47
(14,7,3)	7.43	0.25	15.67	5.34	4.16	7.08
MLP architectures with 2 hidden layers						
Architecture	Cross-validation MSE on the test data [$\cdot 10^{-3}$]			Cross-validation MSE on the training data [$\cdot 10^{-3}$]		
	mean	10%-ile	90%-ile	mean	10%-ile	90%-ile
(14,10,8,3)	7.16	0.17	16.38	4.97	3.40	7.11
(14,6,6,3)	7.22	0.21	15.47	5.03	3.82	6.75
(14,8,8,3)	7.66	0.17	15.16	4.66	3.54	6.26
(14,9,9,3)	7.68	0.21	17.14	4.68	3.39	6.25
(14,11,11,3)	7.72	0.21	17.09	4.68	3.39	6.20

1. Each MLP was employed to approximate the conversions of CH_4 and NH_3 and the yield of HCN for the 92 materials from the seventh generation of the genetic algorithm.
2. From the two conversions and the yield predicted by those approximations, and from the corresponding measured values, the mean squared error and mean absolute error were calculated for each MLP.
3. For each of the eight architectures with one hidden layer and each of the 21 architectures with two hidden layers with numbers of hidden neurons from the above ranges, a single MLP was trained, using the data about all the 604 catalytic materials considered during the architecture search.

The obtained MSE and MAE values are depicted and compared with the average values of the cross-validation MSE on test data in Figure 8.3 (for MLPs with one hidden layer) and Figure 8.4 (for MLPs with two hidden layers). Figure 8.3 and Figure 8.4 show that errors of the approximation computed by the trained multilayer perceptrons for

Validation of the architecture choice for MLPs with one hidden layer

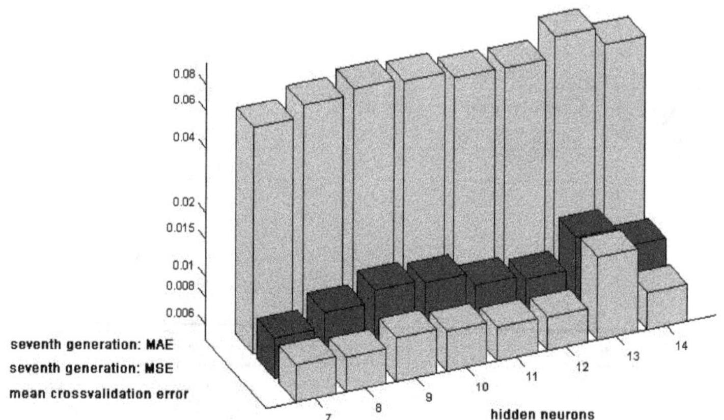

Figure 8.3. Mean absolute errors and mean squared errors of approximations computed for materials from the seventh generation of the genetic algorithm by MLPs with one-hidden-layer architectures fulfilling $7 \leq n_H \leq 14$, trained with all the data considered during the architecture search. For comparison, average cross-validation error values of the involved architectures are recalled from Figure 8.1.

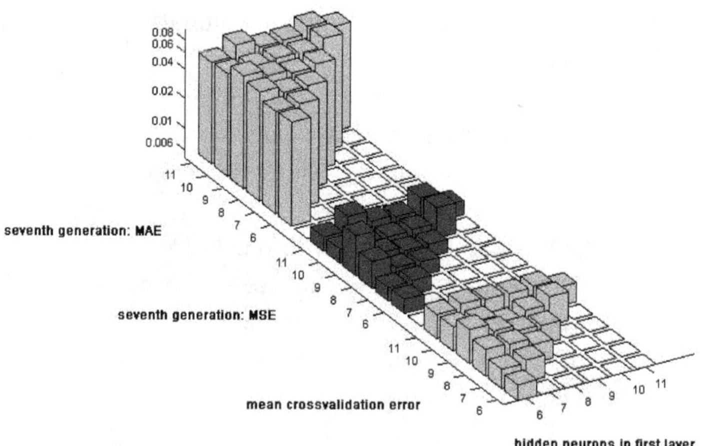

Figure 8.4. Mean absolute errors and mean squared errors of approximations computed for materials from the seventh generation of the genetic algorithm by MLPs with two-hidden-layers architectures fulfilling $6 \leq n_{H2} \leq n_{H1} \leq 11$, trained with all the data considered during the architecture search. For comparison, mean cross-validation error values of the involved architectures are recalled from Figure 8.2.

catalytic materials from the seventh generation of the genetic algorithm were usually lower if the mean cross-validation errors of their architectures were also lower during the architecture search. Similarly, higher mean cross-validation errors of the architectures typically correspond to higher errors of obtained predictions. More generally, the order of both kinds of errors correlates. To quantitatively check this impression, two methods are available:

(i) Calculating the Kendall's order correlation coefficient (Kendall, 1989) between the prediction errors and mean cross-validation errors. This has values between -1 and +1, with +1 attained only if the order of prediction errors and the order of corresponding cross-validation errors are identical, -1 attained only if the order of prediction errors and the order of corresponding cross-validation errors are exactly opposite, and 0 attained if both orders are independent.
(ii) Testing the hypothesis of independence of the order of prediction errors and the order of corresponding cross-validation errors against the alternative of their positive correlation. Then the achieved significance level of the test is a quantitative indication of the strength of that correlation.

We have used both these methods to check the impression given by the above figures. The results are listed in Table 8.2. They clearly confirm that the order of errors of approximations computed with the trained multilayer perceptrons for catalytic materials from the seventh generation and the order of the mean cross-validation errors of their architectures correlate. That correlation is substantially more significant if the errors of approximations are measured in the same way as during the architecture search, i.e., using MSE.

To investigate the usefulness of boosting in our case study, the same data were used and the same set of architectures was considered as for architecture search. In each iteration, a leave-one-out validation was performed, in the manner briefly outlined in the preceding section: the mean squared error of the performance of the catalytic materials serving in the individual folds as test data was calculated, and averaged over all the 604 folds. Boosting was considered useful to those architectures for

Table 8.2. Results of quantitatively checking whether the order of errors of approximations computed by the trained multilayer perceptrons for catalytic materials from the seventh generation of the genetic algorithm correlates with the order of the mean cross-validation errors of their architectures.

Considered networks		one hidden layer (Figure 8.3)	two hidden layers (Figure 8.4)
Kendall's order correlation coefficient	MSE	+0.71	+0.77
	MAE	+0.64	+0.36
Achieved significance for order independence	MSE	$7.1 \cdot 10^{-3}$	$1.7 \cdot 10^{-8}$
	MAE	$1.6 \cdot 10^{-2}$	$1.1 \cdot 10^{-2}$

which the average MSE was in the second iteration lower than in the first iteration. Moreover, the iteration until which the average MSE continuously decreased was then taken as the *final iteration of boosting*.

According to the above criterion, boosting was useful to nine out of the 12 considered architectures with one hidden layer and to 65 out of the 78 considered architectures with two hidden layers. For those architectures, basic information relevant to boosting is summarised in Table 8.3. Notice, in particular, that boosting was useful neither for the architecture (14,12,3), which was among the considered architectures with one hidden layer the best from the point of view of the cross-validation MSE on the test data, nor for the architecture (14,10,8,3), which was from that point of view the best among the considered architectures with two hidden layers.

To validate the most promising results of the investigation of the usefulness of boosting in our case study, again the data from the seventh generation of the genetic algorithm were used. The validation included the *five architectures* that were most promising for boosting from the point of view of the *lowest boosting MSE on the test data in the final iteration*. According to Table 8.3, these were the architectures (14,10,6,3), (14,14,8,3), (14,13,5,3), (14,10,4,3) and (14,11,3). For each of them, the validation proceeded as follows:

Table 8.3. Architectures for which boosting improved the average MSE on the test data, together with the final iteration until which it was improved, and its values in the first and final iteration.

Architecture	Final iteration	Boosting MSE on the test data [$\cdot 10^{-3}$]		Architecture	Final iteration	Boosting MSE on the test data [$\cdot 10^{-3}$]	
		First iteration	Final iteration			First iteration	Final iteration
(14,3,3)	2	8.89	8.88	(14,10,10,3)	3	9.70	9.41
(14,4,3)	7	8.99	8.86	(14,11,3,3)	7	11.30	9.03
(14,5,3)	3	8.17	7.11	(14,11,5,3)	8	9.81	7.98
(14,6,3)	3	7.80	7.75	(14,11,6,3)	6	8.69	7.82
(14,8,3)	2	7.22	7.20	(14,11,7,3)	7	9.59	8.92
(14,9,3)	2	7.95	7.92	(14,11,8,3)	24	9.18	7.53
(14,11,3)	3	7.13	7.11	(14,11,10,3)	2	9.28	9.13
(14,13,3)	2	12.11	12.03	(14,12,3,3)	11	10.94	7.31
(14,14,3)	3	7.38	7.35	(14,12,4,3)	13	10.80	8.46
(14,3,3,3)	2	8.78	8.67	(14,12,5,3)	8	8.97	8.32
(14,4,3,3)	19	8.74	7.50	(14,12,6,3)	6	9.18	8.71
(14,4,4,3)	11	9.25	8.67	(14,12,7,3)	7	10.11	9.17
(14,5,3,3)	5	8.94	8.16	(14,10,10,3)	3	9.70	9.41
(14,5,5,3)	3	8.28	7.92	(14,11,3,3)	7	11.30	9.03
(14,6,3,3)	8	9.16	8.46	(14,11,5,3)	8	9.81	7.98
(14,6,4,3)	3	8.55	8.22	(14,11,6,3)	6	8.69	7.82
(14,6,6,3)	3	7.22	7.16	(14,11,7,3)	7	9.59	8.92
(14,7,3,3)	5	8.54	7.70	(14,11,8,3)	24	9.18	7.53
(14,7,4,3)	5	9.24	8.22	(14,11,10,3)	2	9.28	9.13
(14,7,5,3)	2	8.58	8.29	(14,12,8,3)	6	8.13	7.85
(14,7,6,3)	5	7.75	7.56	(14,12,9,3)	7	9.91	9.10
(14,7,7,3)	4	8.27	7.27	(14,12,10,3)	2	8.47	8.40
(14,8,3,3)	17	10.47	8.18	(14,12,12,3)	13	8.62	7.31
(14,8,4,3)	6	11.06	9.96	(14,13,3,3)	9	11.43	9.30
(14,8,5,3)	5	8.54	7.91	(14,13,4,3)	2	10.08	8.70
(14,8,6,3)	14	9.23	8.34	(14,13,5,3)	31	9.55	6.74
(14,8,8,3)	2	7.66	7.53	(14,13,6,3)	20	9.55	7.27
(14,9,3,3)	8	10.24	7.60	(14,13,7,3)	7	9.29	8.59
(14,9,4,3)	2	9.06	8.58	(14,13,8,3)	3	8.68	8.14
(14,9,5,3)	10	8.96	8.13	(14,13,9,3)	8	9.52	8.19
(14,9,6,3)	8	10.94	10.08	(14,13,10,3)	3	8.44	8.22
(14,9,7,3)	5	9.09	8.74	(14,13,12,3)	13	9.08	7.60
(14,10,3,3)	8	10.08	9.00	(14,13,13,3)	13	8.39	7.41
(14,10,4,3)	19	9.63	7.07	(14,14,3,3)	6	11.52	8.89
(14,10,5,3)	3	8.78	8.35	(14,14,4,3)	15	9.56	7.13
(14,10,6,3)	32	8.69	6.55	(14,14,5,3)	7	10.43	8.82
(14,10,7,3)	4	8.63	8.04	(14,14,7,3)	7	9.04	7.87

1. In each iteration up to the final boosting iteration corresponding to the respective architecture according to the above investigation (Table 8.3), a single MLP was trained with data about all the 604 catalytic materials considered during the architecture search.
2. Each of those MLPs was employed to approximate the conversions of CH_4 and NH_3 and the yield of HCN for the 92 materials from the seventh generation of the genetic algorithm.
3. In each iteration, the medians with respect to the probability distribution (12) of the approximations of the two conversions and of the HCN yield obtained up to that iteration were used as their boosting approximations.
4. From the conversions and the yield predicted by the boosting approximations, and from the measured values, the boosting MSE and boosting MAE were calculated for each MLP.

The resulting boosting errors (MSE and MAE) are summarised in Figure 8.5, whereas Figure 8.6 compares the boosting approximations of the conversions of CH_4 and NH_3 and of the yield of HCN in the first and final iteration with their measured values.

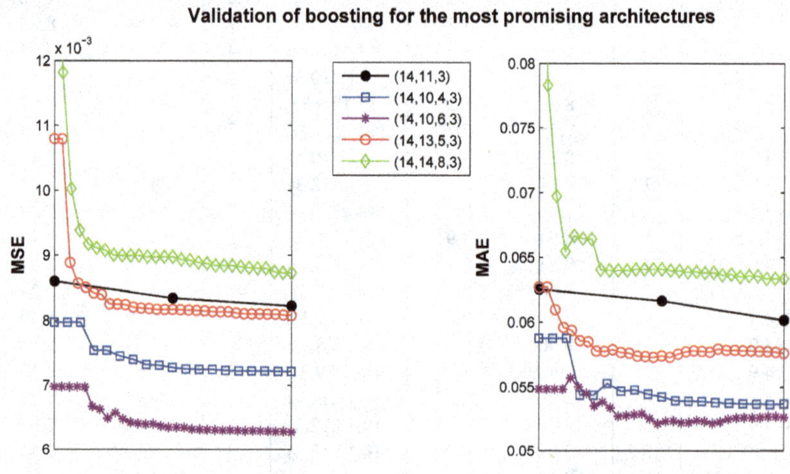

Figure 8.5. History of the boosting MSE and MAE on the data from the seventh generation of the genetic algorithm for MLPs with the five architectures included in the validation of boosting.

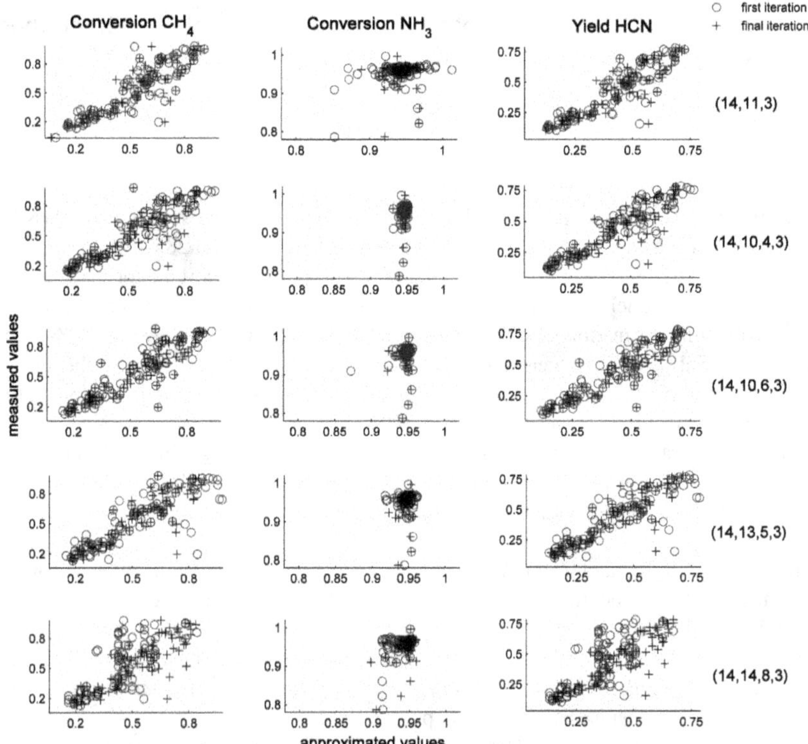

Figure 8.6. Comparison of the boosting approximations of the conversions of CH_4 and NH_3 and of the yield of HCN in the first and final iteration with their measured values for the 92 catalytic materials from the seventh generation of the genetic algorithm. The approximations were obtained using the MLPs from Figure 8.5.

The results presented clearly confirm the usefulness of boosting for the five considered architectures. For each of them, boosting leads to an overall decrease of both considered error measures, MSE and MAE, on new data from the seventh generation of the genetic algorithm. Moreover, the decrease of the MSE (which is the measure employed during the investigation of the usefulness of boosting) is uninterrupted or nearly uninterrupted until the final boosting iteration. On the other hand, the scatter plots in Figure 8.6 do not indicate any apparent difference between the effect of boosting on the three properties employed as catalyst performance measures in our case study — conversion of CH_4, conversion of NH_3, and yield of HCN. Hence, the performed validation confirms the usefulness of boosting irrespectively of which of those performance measures is considered.

Bibliography

Altinçay, H. (2004). Optimal resampling and classifier prototype selection in classifier ensembles using genetic algorithms, *Pattern Anal. Applic.*, 7, 285–295.

Baumes, L.A., Serra, J.M., Serna, P. and Corma, A. (2006). Support vector machines for predictive modelling in heterogeneous catalysis: A comprehensive introduction and overfitting estimation base on two real applications, *J. Comb. Chem.*, 8, 583–596.

Corma, A., Serra, J.M., Argente, E., Botti, V. and Valero, S. (2002). Application of artificial neural networks to combinatorial catalysis: Modeling and predicting ODHE catalysts, *ChemPhysChem*, 3, 939–945.

Corma, A., Serra, J.M, Serna, P. and Moliner, M. (2005). Integrating high-throughput characterization into combinatorial heterogeneous catalysis: Unsupervised construction of quantitative structure/property relationship models, *J. Catal.*, 232, 335–341.

Cundari, T., Deng, J. and Zhao, Y. (2001). Design of a propane ammoxidation catalyst using artificial neural networks and genetic algorithms, *Ind. Eng. Chem. Res.*, 40, 5475–5480.

Drucker, H. (1997). Improving regression using boosting techniques. In: Fisher, D.H. (ed.), *Proceedings of the 14th International Conference on Machine Learning*, Morgan Kaufmann, San Francisco, pp. 107–115.

Freund, Y. and Schapire, R. (1997). A decision-theoretic generalization of on-line learning and an application to boosting, *J. Comput. System Sci.*, 55, 119–139.

Friedman, J. (2001). Greedy function approximation: A gradient boosting machine, *Ann. Statist.*, 29, 1189–1232.

Friedman, J., Hastie, T. and Tibshirani, R. (2000). Additive logistic regression: A statistical view of boosting, *Ann. Statist.*, 28, 337–374.

Günay, M.E. and Yildirim, R. (2008). Neural network aided deisgn of Pt-Co-Ce/Al_2O_3 catalyst for selective CO oxidation in hydrogen-rich streams, *Chem. Eng. J.*, 140, 324–331.

Hand, D.J. (1997). *Construction and Assessment of Classification Rules*, Wiley, New York, 214 p.

Hoffmann F. (2004). Combining boosting and evolutionary algorithms for learning of fuzzy classification rules, *Fuzzy Sets Syst.*, 141, 47–58.

Hou, Z.Y., Dai, Q., Wu, X.Q. and Chen, G.T. (1997). Artificial neural network-aided design of catalyst for propane ammoxidation. *Appl. Catal., A: General*, 161, 183–190.

Huang, K., Feng-Qiu, C. and Lü, D.W. (2001). Artificial neural network-aided design of a multi-component catalyst for methane oxidative coupling, *Appl. Catal., A: General*, 219, 61–68.

Kendall, D.G. (1989). A survey of the statistical theory of shape, *Stat. Sci.*, 4, 87–99.

Kito, S., Hattori, T. and Murakami, Y. (1994). Estimation of catalytic performance by neural network — Product distribution in oxidative dehydrogenation of ethylbenzene, *Appl. Catal., A: General*, 114, L173–L178.

Redpath, D.B. and Lebart, K. (2005). Boosting feature selection. In: Sing, S. *et al.* (eds.), *Pattern Recognition and Data Mining*, Springer, Berlin, pp. 305–314.

Schapire, R. (1990). The strength of weak learnability, *Mach. Learning*, 5, 197–227.

Shrestha, D.L. (2005). Experiments with AdaBoost.RT, an improved boosting scheme for regression, *Neural Comput.*, 18, 1678–1710.

Tompos, A., Margitfalvi, J.L., Tfirst, E. and Végvári, L. (2003). Information mining using artificial neural networks and "holographic research strategy", *Appl. Catal., A: General*, 254, 161–168.

Tompos, A., Margitfalvi, J.L., Tfirst, E. and Végvári, L. (2006). Evaluation of catalyst library optimization algorithms: Comparison of the holographic research strategy and the genetic algorithm in virtual catalytic experiments, *Appl. Catal., A: General*, 303, 72–80.

White, H. (1992). *Artificial Neural Networks: Approximation and Learning Theory*, Blackwell, Cambridge, 320 p.

Zemel, R. and Pitassi, T. (2001). A gradient-based boosting algorithm for regression problems. In: Leen, T.K., Dietterich, T.G. and Tresp, V. (eds.), *Advances in Neural Information Processing Systems 13*, MIT Press, Cambridge (MA), pp. 696–702.

Chapter 9

Applications of Combinatorial Catalyst Development and an Outlook on Future Work

9.1. Introduction

Selected applications of the different mathematical methods have already been explained in the form of case studies in the preceding Chapters 5, 7 and 8.

Section 5.3: Analysis and mining of data collected in catalytic experiments: HCN synthesis.

Section 7.3: Tuning evolutionary algorithms with artificial neural networks: Oxidative dehydrogenation of propane.

Section 8.4: Improving neural network approximations: HCN synthesis.

Progress in combinatorial development of heterogeneous catalysts during the last five years is summed up in the following illustrations for the years 2003 to early 2009 (Section 9.2: experimental development of catalysts; Section 9.3: new methodologies). Reference is made to the various sources where the selected examples are comprehensively described. The selected descriptions are based on the abstracts of the respective publications. In the final Section 9.4 some conclusions are drawn from the work up to now and an outlook for further development in the field is presented.

9.2. Experimental Applications of Combinatorial Catalyst Development

Example 9.2-1 Multi-component Au/MgO catalysts designed for selective oxidation of carbon monoxide: Application of a combinatorial approach (Tompos *et al.*, 2008).

A complex combinatorial approach was applied for the design of multicomponent Au/MgO catalysts for CO oxidation in the presence of hydrogen. The combinatorial library design led to unique and previously unknown catalyst compositions. The best catalysts contained significant amounts of Pb and Sm, which had never before been reported as useful components of PROX catalysts. In order to increase the diversity of the experimental space, different pretreatment conditions were applied prior to catalytic tests. Actually, two catalyst libraries were tested after (i) a reductive, and (ii) a combined reductive pretreatment. Different optimum compositions have been obtained after using these two pretreatment procedures. The combined reductive treatment resulted in a more cost-effective optimum composition with significantly fewer components and smaller gold content in comparison to the best catalyst obtained after using a simple reductive treatment. Further studies showed different ways of promoting the action of modifiers. After reductive pretreatment the modifiers suppress hydrogen consumption, while after combined reductive pretreatment the modifiers lead to the promotion of both oxidation reactions.

Example 9.2-2 New catalytic materials for the high-temperature synthesis of hydrocyanic acid from methane and ammonia by a high-throughput approach (Moehmel *et al.*, 2008).

For converting methane and ammonia to hydrocyanic acid, catalysts were prepared and tested in a 48-parallel-channel fixed-bed reactor unit operating at temperatures up to 1,373K. The catalysts were synthesised using a robot applying a genetic algorithm as the design tool. New and improved catalyst compositions were discovered by using a total of seven generations each consisting of 92 potential catalysts. The catalyst support turned out to be an important input variable. Furthermore, platinum, which is well known as a catalytic material, was confirmed. Moreover, improvements in HCN yield were achieved by addition of promoters like Ir, An, Ni, Mo, Zn and Re. Multi-way analysis of variance and regression trees were applied to establish correlations between HCN yield and catalyst composition (support and metal additives). The results obtained may be considered as the base for future even more efficient screening experiments.

Example 9.2-3 A combinatorial approach for the discovery of low-temperature soot oxidation catalysts (Olong *et al.*, 2007).

High-throughput syntheses using a commercial pipetting robot with the aid of the 'Plattenbau' software as well as a highly parallel screening technique were used in the search for low temperature soot oxidation catalysts. It has been shown that emissivity-corrected infrared thermography which monitors the heat changes resulting from the heat of reaction on catalyst surfaces is an efficient and fast screening technique in soot oxidation catalysts. It was found that alkali metal mixed oxides have the potential to decrease the soot oxidation temperature to diesel exhaust realistic temperatures. From the first generation, the oxides K3Ce97 and Cs3Co97 were found to be the most active and stable samples. Further doping and composition spread of the CsCo catalyst resulted in samples that are even more active. Hits of HT-screening experiments were successfully confirmed by thermogravimetric analysis.

Example 9.2-4 Discovery of new heterogeneous catalysts for the selective oxidation of propane to acrolein (Oh *et al.*, 2007).

Combinatorial synthesis and screening technique were applied to investigate the catalytic activity and selectivity of ternary and quaternary mixed-metal oxide catalysts for the selective oxidation of propane. The catalyst libraries were prepared via a modified sol-gel method using a synthesis robot and library design software, and examined for the catalytic activities in a simple high-throughput reactor system connected to a mass spectrometer for product analysis. Ternary Mo-Cr-Te, V-Cr-Sb, and Mo-V-Cr catalysts were selected as potential candidates by a composition-spread approach. In the next-generation of a composition-spread library, the composition space of these three ternary compositions was sampled. Screening of this 198-member library provided substantial evidence that each ternary system had its own optimum composition where acrolein formation was highest. In addition, the composition space of the quaternary reference system Mo-V-Te-Nb mixed-oxides was prepared and sampled.

Example 9.2-5 Characterization of trimetallic Pt-Pd-Au/CeO_2 catalysts designed for methane total oxidation by combinatorial methods (Tompos *et al.*, 2007).

The role and effect of platinum and gold on the catalytic performance of ceria(CeO_2)-supported tri-metallic Pt-Pd-Au catalysts were studied. An optimum composition of these tri-metallic supported catalysts was discovered using methods and tools of combinatorial catalyst library design. Detailed catalytic, spectroscopic and physico-chemical characterisations of catalysts in the vicinity of the optimum in the given compositional space were performed. The temperature-programmed oxidation of methane revealed that the addition of Pt and Au to Pd/CeO_2 catalyst resulted in higher conversion values in the whole investigated temperature range compared to the monometallic Pd catalyst. The time-on-stream experiments provided further evidence for the high-stability of tri-metallic catalysts compared to the monometallic one. Kinetic studies revealed the stronger adsorption of methane on Pt-Pd/CeO2 catalysts than over Pd/CeO_2. XPS analysis showed that Pt and Au stabilise Pd in a more reduced form even under condition of methane oxidation. FTIR spectroscopy of adsorbed CO and hydrogen TPD measurements provided indirect evidence for alloying of Pt and An with Pd. CO chemisorption data indicated that tri-metallic catalysts reaching an optimum ratio between Pd-0 and PdO species, and (iii) stabilization of Pd in high dispersion. The results also indicate that Pd-0-PdO ensemble sites required for methane activation have increased accessible metallic surface area. It is suggested that advantageous catalytic properties of tri-metallic Pt-Au-Pd/CeO_2 catalysts compared to the monometallic one can be attributed to (i) suppression of the formation of ionic forms of Pd(H), (ii) reaching an optimum ratio between Pd-0 and PdO species, and (iii) stabilisation of Pd in high dispersion. The results also indicate that Pd-0-PdO ensemble sites are required for methane activation.

Example 9.2-6 Utilization of combinatorial method and high throughput experimentation for development of heterogeneous catalysts (Yamada and Kobayashi, 2006).

The combinatorial method and high-throughput experimentation have been applied to catalysis development in the last decade. Here equipment for catalyst preparation, catalysis evaluation and product analysis with a gas-sensor system is described. High-throughput experimentation using these devices allowed the researchers to optimise catalyst composition with stochastic methods such as the genetic algorithm. The improvement of propane selective oxidation catalysis is described. High-throughput experimentation also allowed the construction of a database for elemental reactions. The investigators developed novel catalysts for ethanol steam-reforming catalysts and a dimethyl ether steam-reforming catalyst by combining good catalysts for each elemental reaction. The new concept of 'MATERIOMICS' is introduced as a promising method for material science based on combinatorial technology.

Example 9.2-7 Discovery of new catalytic materials for the water-gas shift reaction (WGSR) by high-throughput experimentation (Grubert *et al.*, 2006).

New Cu-free non-pyrophoric catalytic materials have been discovered for the WGSR by high-throughput experimentation applying an evolutionary search strategy. Two approaches were applied for the design of experiments: (1) the amount of Cu in the composition was not restricted (as expected from common knowledge Cu containing catalysts were superior within the given parameter space), (2) by allowing only a maximum Cu content of 1 wt.% in the second approach, new Cu-free materials were discovered. The new compositions contained mainly oxides of Cr or Fe along with Mn and Pt on ZrO_2 as support material. A maximal activity for WGSR was achieved at 250°C (CO conversion of 55%); the feed composition amounted to 3% CO, 37% H_2, 14%CO_2, 23% H_2O, Ar balance and the GSHV to 3000 h^{-1}.

Example 9.2-8 Optimisation of olefin epoxidation catalysts applying high-throughput techniques and genetic algorithms assisted by artificial neural networks (Corma *et al.*, 2005).

An olefin epoxidation Ti-silicate-based catalyst was optimised by means of high-throughput experimentation for catalytic-materials synthesis, postsynthesis treatments, and catalytic testing. Soft-computing techniques for advanced experimental design and data assessment (GA, ANN) were used. The following variables were explored in the hydrothermal synthesis of Ti-silicate-based catalysts: concentration of OH^-, titanium, and surfactant. The probe reaction employed for the optimisation was the solvent-free epoxidation of cyclohexene with *tert*-butylhydroperoxide as oxidant. The different groups of catalysts were subjected to a clustering analysis and studied by various spectroscopic means. Ti-mesoporous MCM-41 and MCM-48 molecular sieves were among the most active catalysts, which were also tested for epoxidation of different linear olefins.

Example 9.2-9 Optimisation of MoVSb oxide catalyst for partial oxidation of isobutane by combinatorial approaches (Paul *et al.*, 2005).

Optimisation of the Mo-V-Sb mixed-oxide system for the selective oxidation of isobutane to methacrolein by combinatorial methods is primarily intended to reduce the number of experiments in a broad parameter space. Therefore, an evolutionary approach based on a genetic algorithm has been chosen to screen three generations of 30 catalysts. With the help of automated sol-gel synthesis techniques, a high-throughput continuous flow reactor (16UPCFR) and appropriate software for experimental design, a new catalyst composition with improved performance has been obtained. Finally, the best catalysts were scaled up to gram quantities and tested in a continuous-flow reactor unit that was equipped with four parallel reactors (4UPCFR). The final catalyst showed a significantly higher selectivity toward methacrolein at the same isobutane conversion, compared to the initial $Mo_8V_2Sb_{90}O_x$ catalyst.

Example 9.2-10 Optimisation of Cu oxide catalysts for methanol synthesis by combinatorial tools (Watanabe *et al.*, 2004).

For methanol synthesis from syngas, catalysts of high activity at low pressure and low temperature are needed. The combinatorial approach comprised a high-pressure high-throughput screening multi-reactor setup (microplates for handling 96 catalytic materials simultaneously). Results from activity tests were used for training an ANN. Then, catalyst activity was mapped by ANN as a function of catalyst composition comprising Cu, Zn, Al, Sc, B and Zr and parameters of catalyst preparation. The composition of the optimum catalyst was Cu(43), Zn(17), Al(23), Sc(11), B(0), Zr(6); its activity (427g methanol/kg catalyst and h) was much higher than that of an industrial catalyst (250 g methanol/kg catalyst and h) at 1MPa and 498K.

Example 9.2-11 Directed evolution of noble-metal free catalysts for the oxidation of CO at room temperature (Saalfrank and Maier, 2004).

IR-thermography and combinatorial library design (doping and composition spread) lead in a few generations to new, noble-metal-free catalysts for the low-temperature oxidation of CO in air. Activity patterns could be derived from the emissivity-corrected IR thermographic image of a catalyst library.

Example 9.2-12 Development of low-temperature light-paraffin isomerisation catalysts with improved resistance to water and sulfur by combinatorial methods (Serra *et al.*, 2003).

For the search of new more-thioresistant catalysts for low-temperature isomerisation of light paraffins, combinatorial techniques (high-throughput catalyst preparation and testing, and a GA) were used. After three iterative cycles (generations) catalysts have been found that were not only active and selective but also more resistant to deactivation by water and sulphur than the corresponding conventional ones. The results were reproducible in a pilot plant.

Example 9.2-13 Catalyst design for oxidative methane coupling by using an artificial neural network and a hybrid genetic algorithm (Huang *et al.*, 2003).

A multi-component catalyst for the title reaction was optimised with a hybrid GA achieving ethane + ethylene yields (C_2) larger than 25% at 1,069°C. In a specific case, methane conversion amounted to 37.8% and C_2 selectivity to 73.5%, i.e. C_2 yield was 27.8% obtained. An artificial neural network described the relationship between catalyst composition and catalyst performance. The results indicate a prospect for industrialisation.

Example 9.2-14 Catalyst optimisation for methanol synthesis by a combinatorial approach using a genetic algorithm assisted by a neural network (Umegaki *et al.*, 2003).

For methanol from synthesis gas, catalyst compositions based on Cu-Zn-Al-Sc oxides and calcination temperature of the catalyst precursor were optimised for achieving maximum activity. A combination of GA and a radial basis function network turned out to be more robust than GA alone.

Example 9.2-15 Combinatorial and conventional development of novel dehydrogenation catalysts (Urschey *et al.*, 2003).

A combinatorial search for novel catalysts for the oxidative dehydrogenation of *n*-butane is presented. A total of 950 mixed oxides based on SiO_2, TiO_2, ZrO_2 and Al_2O_3 were prepared on a milligram scale using a pipetting robot. These materials have been screened for catalytic activity for the ODH reaction of ethylbenzene and *n*-butane by means of emissivity-corrected IR thermography. Selected active materials have been synthesised on a gram scale and tested in a conventional fixed-bed reactor. A parameter optimisation study using a factorial design and data visualisation software revealed several regions of operability, where high selectivities and good yields were obtained. The best new catalyst found, $Hf_3Y_3Ti_{94}O_x$, reached a butene selectivity of 75% at 15% butane conversion at 450°C and ambient pressure, which compares well with state-of-the-art catalysts for this reaction.

Example 9.2-16 Design of propane ammoxidation catalysts using ANNs and GAs (Cundar *et al.*, 2001).

The target of this work was the optimal design of a catalyst for the ammoxidation of propane using ANN and GA. Serving as input for the combinatorial approach were the oxides of P, K, Cr, Mo, Al, I, V, Sb, W, and Sn, and serving as output (objective function) were the activity and the selectivity/yield of acrylonitrile. A trained optimal linear-combination (OLC) neural network was used to correlate input and output data obtained during the optimisation procedure applying GA. The best yield of acrylonitrile after GA procedure was 79%, which was higher than the highest yield previously reported at that time (64%). The OLC neural network, using the acrylonitrile yield as output, greatly improved the simulation of the composition of the catalytic material compared to a simple, single-network architecture. In particular, whereas single-network methods can easily reproduce the experimental patterns used for training and validation. The OLC-ANN is markedly superior for generalising performance patterns of novel catalyst compositions. — In terms of neural network training as explained in the current monograph (Chapter 6, Section 6.3), the above abstract indicates that the OLC neural networks were less prone to overtraining than fully connected MLPs, and therefore better in generalising for novel catalyst compositions.

9.3. Methodology

Example 9.3-1 DoE framework for catalyst development based on soft computing techniques (Valero *et al.*, 2009).

Soft computing techniques have recently been successfully applied in the chemical engineering field for reducing both high temporal costs and financial costs. A generalistic, configurable and parametrisable experimental design framework has been developed for the discovery and optimisation of catalytic materials when exploring a high-dimensional space. This framework is based on a soft computing architecture in which neural networks and a genetic algorithm are combined to optimise the discovery of new materials and process conditions in catalytic reactors on an industrial scale. Users can configure the parameters of the soft computing technique, adapting it to the specific features of each experiment. Moreover, a new problem codification for genetic algorithms has been developed, which deals with optimisations simultaneously considering complex catalyst formulations and different synthesis/testing conditions as variables. — The proposed framework has been employed in several examples, in which both the developed codification and the framework parameterization have been tested.

Example 9.3-2 On the suitability of different representations of solid catalysts for combinatorial library design by genetic algorithms (Gobin and Schueth, 2008).

Genetic algorithms are widely used to solve and optimise combinatorial problems and are more often applied for library design in combinatorial chemistry. Because of their flexibility, however, their implementation can be challenging. In this study, the influence of the representation of solid catalysts on the performance of genetic algorithms was systematically investigated on the basis of a new, constrained, multiobjective, combinatorial test problem with properties common to problems in combinatorial materials science. Constraints were satisfied by penalty functions, repair algorithms, or special representations. Performing 100 optimisation runs for each algorithm and test case carried out the tests using three state-of-the-art evolutionary multiobjective algorithms. Experimental data obtained during the optimisation of a noble metal-free solid catalyst system active in the selective catalytic reduction of nitric oxide with propene was used to build up a predictive model to validate the results of the theoretical test problem. A significant influence of the representation on the optimisation performance was observed. Binary codes were mostly found to be the preferred ones, and depending on the experimental test unit, repair algorithms or penalty functions performed best.

Example 9.3-3 Array-based split-pool combinatorial screening of potential catalysts (Stanton, 2007).

A new method for screening split-pool combinatorial libraries for catalytic activity is described. Site-selective detection of catalytic activity for solution-based reactions was made possible without co-functionalising beads or adding diffusion-limiting matrices. This was done by spatially separating resin-bound catalysts on an adhesive array on a microscope slide and introducing the reacting liquid to the top of the slide. Convective mixing and evaporation was controlled using a cover slide and imaging both the formation of products within active beads and the diffusion of products out of the beads. Coloured reaction products and pH-sensitive indicators were used to visually detect catalytically active beads in the presence of inactive ones. Quantitative analyses of the images support the assumption that colour intensities can be used to assess the quality of hits from a combinatorial screen. The Knoevenagel condensation reaction catalysis, as well as esterase screening using methyl red, were used to validate the approach. Using the esterase data, it was shown that some information on activity could also be extracted from the coloured plume surrounding individual beads, although the precision is not as good as that from direct measurement of absorbance through the bead. It was also found that the distribution of products within a single bead can be gleaned from the absorbance data for different-sized beads.

Example 9.3-4 Diversity management for efficient combinatorial optimisation of materials (Farrusseng and Clerc, 2007).

The issues of evolutionary library design in the frame of material and catalyst discovery were described. Concepts of diversity management on material library to enhance the efficiency of the optimisation were proposed. The diversity monitoring is implemented via two different approaches. The first deals with a dynamic monitoring of mutation and crossover rates, whereas the second involves a selection step based on sample 'distance'. Simulations of optimisation were performed on a surface response, which was designed to mimic realistic data. Algorithm performances were compared in terms of both efficiency and reliability.

Example 9.3-5 Library design using genetic algorithms for catalyst discovery and optimisation, Review of Scientific Instruments (Clerc *et al.*, 2005).

A detailed investigation of catalyst library design by genetic algorithm (GA) is reported. A methodology for assessing GA configurations is described. Operators, which promote the optimisation speed while being robust to noise and outliers, are revealed through statistical studies. The genetic algorithms were implemented in GA platform software called OptiCat, which enables the construction of custom-made workflows using a toolbox of operators. Two separate studies were carried out, (i) on a virtual benchmark and (ii) on real surface response derived from HT screening. Additionally, a methodology is described to model a complex surface response by binning the search space in small zones, which are then independently modelled by linear regression. In contrast to artificial neural networks, this approach allows one to obtain an explicit model in an analogical form that can be further used in Excel or entered in OptiCat to perform simulations. Speeding up the implementation of a hybrid algorithm via combining a GA with a knowledge-based extraction engine the optimisation process is accelerated by virtual prescreening. In this way, the hybrid GA opens the 'black-box' by providing knowledge as a set of association rules.

Example 9.3-6 Integrating high-throughput characterization into combinatorial heterogeneous catalysis: unsupervised construction of quantitative structure/property relationship models (Corma *et al.*, 2005).

A novel approach is presented in the framework of heterogeneous combinatorial catalysis, which integrates into the global discovery strategy the use of inexpensive high-throughput characterisation of libraries of catalysts, as multivariate spectral descriptors for catalytic quantitative structure/property relationship (QSPR) modelling. Moreover, QSPR models can be used to aid the design of new libraries and for extraction of rules and relationships, yielding knowledge about catalysis. This approach can be of special interest when experimental evaluation of catalytic behaviour is very expensive or time-consuming, — as, for instance, in catalyst deactivation studies, in testing under very severe conditions, or when large amounts of catalyst are demanded —. This methodology has been applied to modelling of the behaviour of epoxidation catalysts, with the composition vector of the starting synthesis gel and XRD spectra as descriptors. Dimensional reduction was conducted by principal components analysis, clustering, and Kohonen networks; predictive models were obtained with the use of logistic equations, artificial neural networks, and decision-tree techniques. The use of spectral descriptors made it possible to markedly improve the prediction of performance obtained with synthesis descriptors alone.

Example 9.3-7 Spatially resolved mass spectrometry as a fast semi-quantitative tool for testing heterogeneous catalyst libraries under reducing stagnant-point flow conditions (Eckhard *et al.*, 2005).

A calibrated quadrupole mass spectrometer (QMS) is combined with a homemade positioning unit for applying scanning electrochemical microscopy, which sequentially addresses 25 catalysts, which are placed in wells. Highly reproducible catalytic data are obtained under stagnant-point flow conditions by means of coupled gas feed and QMS capillaries. The reaction array can be heated and is fully sealed from the atmosphere by a glass lid. Additionally, the reaction chamber is flushed with argon making it possible to study the structure-insensitive hydrogenation of ethene over SiO_2-supported palladium catalysts unimpaired by oxygen poisoning. Since the derived semi-quantitative degree of conversion is proportional to the Pd metal area under fixed reaction conditions, spatially resolved mass spectrometry provides a novel method for rapidly estimating the metal surface areas of a Pd catalyst library.

Example 9.3-8 Efficient discovery of nonlinear dependencies in a combinatorial catalyst data set (Cawse *et al.,* 2004).

Exploration of a complex catalyst system using genetic algorithm methods and combinatorial experimentation efficiently removes non-contributing elements and generates data that can be used to model the remaining system. In particular, the combined methods effectively navigate and optimise systems with highly nonlinear dependencies (three-way and higher interactions).

Example 9.3-9 Application of a genetic algorithm and a neural network for the discovery and optimisation of new solid catalytic materials (Rodemerck *et al.*, 2004).

In the process of discovering new catalytic compositions by combinatorial methods in heterogeneous catalysis, usually various potential catalytic compounds have to be prepared and tested. To decrease the number of necessary experiments an optimisation algorithm based on a genetic algorithm for deriving subsequent generations from the performance of the members of the preceding generation is described. This procedure is supplemented by using an artificial neural network for establishing relationships between catalyst compositions or — speaking more generally — i.e., materials properties and their catalytic performance. By combining a trained neural network with the genetic algorithm software, virtual computer experiments were carried out that aimed at adjusting the control parameters of the optimisation algorithm to the special requirement of catalyst development. The approach is illustrated by the search for new catalytic compositions for the oxidative dehydrogenation of propane.

Example 9.3-10 Holographic research strategy for catalyst library design: Description of a new powerful optimisation method (Vegvary *et al.*, 2003).

The principle of a deterministic optimisation algorithm called holographic research strategy (HRS) is described. It was demonstrated that in a multi-dimensional experimental space having around 63,000 potential experimental points fewer than 200 virtual experiments were sufficient to find the global maximum. The benefits of the use of HRS are as follows: (i) easy application to catalyst library design, (ii) speed in finding the global optimum, and (iii) good visualisation of a multidimensional experimental space.

Example 9.3-11 High-throughput experimental and theoretical predictive screening of materials — A comparative study of search strategies for new fuel cell anode catalysts (Strasser *et al.*, 2003).

A comparative study of experimental and theoretical combinatorial and high-throughput screening methods for the development of novel materials is presented. Both methods were applied to the development of new anode fuel-cell alloy catalysts with improved CO tolerance. Combinatorial electrocatalysis experiments were performed on a 64-element electrode array. Sputter-deposited ternary thin-film electrocatalysts of composition PtRuM (M = Co, Ni or W) were screened in parallel for their methanol oxidation activity, and their individual geometric and specific chronoamperometric current density were monitored and evaluated against standard PtRu catalysts. Density functional theory calculations of a variety of model ternary PtRuM alloy catalysts yielded detailed adsorption energies and activation barriers. Feeding these thermodynamic and kinetic data into a simple micro-kinetic model for the CO electro-oxidation reaction, the relative activities of a number of PtRuM ternary alloys were calculated. The experimental and theoretical computational results reveal very similar trends in electro-catalytic activity as a function of alloy composition; they also point at similar ternary PtRuM alloys as candidates for improved anode catalysts for low-temperature fuel cells.

9.4. Conclusions and Outlook

9.4.1. *Applications of Combinatorial Methodologies in Practice*

The success of combinatorial methodologies for finding new or improved catalytic materials has been illustrated for a broad spectrum of chemical reactions, as can be seen from the examples summarised in Section 9.2 of this chapter. In some cases significant catalyst improvements as compared to present industrial practice were obtained (cp. examples 9.2-2, 9.2-10, 9.2-12, 9.2-16; see also Baerns *et al.*, 2002).

Different experimental approaches and tools in finding novel catalysts have already been successfully applied; also, different methods of experimental design, data analysis, and data mining, partly called *soft computing* are being used (cp. Section 9.3 of this chapter). There is certainly a challenge for the scientific community to define the optimal approaches for the different types of tasks in combinatorial development.

In most published work, only little information is given about the reproducibility and accuracy of materials preparation and of data accumulated in data assessment during the development process. Their improvement, or at least their rationale with respect to the data on which final evaluation is based, would certainly contribute to a better understanding of and possibly improvement in catalyst development.

Unfortunately, heat and mass transfer limitations in screening catalytic materials have been accounted for only very seldom. In fact, they should be excluded; otherwise, misleading results may be obtained, since there is no comparable basis for data assessment (Baerns and Mirodatos, 2002).

In any advanced combinatorial development, fundamental knowledge ought to be introduced which either already exists or may be gained during the development process (cp. Baerns and Holeňa, 2008). This knowledge primarily comprises surface science processes, i.e., elementary reaction steps required for a complex reaction, and the respective bulk and surface properties of the catalytic material; these properties, which are often called *descriptors,* may include acidity, basicity, redox behaviour, type of electronic conductivity and crystalline structure (cp. Example 9.2.5). Therefore it is suggested that characterisation of the lead compositions generated in the process is beneficial for further optimisation.

The application of fundamental knowledge is, of course, also of particular interest in choosing the primary pool of elements for the first generation of the subsequent combinatorial process; if this is done in an appropriate manner the researcher can also provide for the *unexpected* i.e. catalyst compositions which would not be anticipated by common knowledge or even intuition. Moreover, the large amounts of data that are obtained in high-throughput experimentation could become a most valuable source for deriving fundamental knowledge. This might in the future become a unique opportunity for the science of catalysis by allowing the rapid execution of a large number of experiments aimed at gaining a specific understanding in a specific field of catalysis, which in turn will eventually lead to new perspectives on catalyst development.

High-throughput catalytic testing of solid materials is usually done at standard conditions. However, this should be only considered as a first approximation for identifying suitable materials, catalysing the chemical reaction towards the desired products. To find the optimum conditions for the maximum yield, the whole parameter space of reaction conditions has to be experimentally covered or has to be fully described by the kinetics of the catalytic reaction. The availability of full kinetics for each catalyst has a decisive benefit: Modelling the reaction and simulation of catalyst performance, the maximum yield and selectivity can be computed as a function of reaction conditions for a given type of reactor. In this way, all catalysts can be compared at their individual conditions of maximal performance. On using microkinetics, the effect of heat and mass transport limitations, which might disguise intrinsic operation of the individual catalyst particle, can also be identified.

Even more importantly, if kinetic parameters and catalyst performance can be related to the composition of the catalytic materials, these results may lead to further optimisation of catalyst composition.

9.4.2. *Computer-aided Methods for the Optimisation of Catalyst Composition and Data Mining*

In the area of computer-aided methods for optimisation of catalysts, in particular their composition, attempts can be expected in employing novel kinds of evolutionary optimisation methods, which have already proved to be successful in other application areas. These are, in particular:

- Differential evolution (Feoktistov, 2006).
- Estimation of distribution algorithms (Larranaga and Lozano, 2002).

This development will probably also include other kinds of novel and successful stochastic optimisation methods, especially methods that are, like evolutionary algorithms, based on heuristics inspired by nature. Such methods are in particular:

- Particle swarm optimisation (Kennedy and Eberhart, 2001).
- Ant-colony optimisation (Dorigo and Stützle, 2004).

A similar situation exists in the area of data mining. Nonlinear regression models computed by artificial neural networks are likely to be supplemented by other modern nonlinear regression models, in particular:

- Kernel-based support vector regression (Schölkopf and Smola, 2002).
- Finite mixture models (McLachlan and Peel, 2000).

The support vector approach is actually a more general approach, based on the statistical theory of learning (Vapnik, 1995; Hastie *et al.*, 2001), which pertains also to cluster analysis and especially to classification. The first applications of this approach to catalysis have been recently reported (Baumes *et al.*, 2006; Serra *et al.*, 2007).

Finally, it may be stated that there are many methods available for experimental design as well as for data analysis and mining. These many tools should be compared so as to draw some conclusions for selecting the best ones in the future.

Bibliography

Baerns, M., Buyevskaya, O., Grubert, G. and Rodemerck U. (2001). Combinatorial methodology and its experimental validation by parallel synthesis, testing and characterization of solid catalytic materials. In: Derouane, E.G., Parmon, V., Lemos, F. and Ribiero, F.R. (eds.). *Principles and Methods for Accelerated Catalysts Design and Testing*. pp. 85–100. Springer, Berlin.

Baerns, M. and Mirodatos, C. (2001). Methods and standards of accelerated testing. In: Derouane, E.G., Parmon, V., Lemos, F. and Ribiero, F.R. (eds.) *Principles and Methods for Accelerated Catalysts Design and Testing*. pp. 469–479. Springer, Berlin.

Baumes, L.A., Serra, J.M., Serna, P. and Corma, A. (2006). Support vector machines for predictive modelling in heterogeneous catalysis: A comprehensive introduction and overfitting estimation base on two real applications, J. Comb. Chem., 8, 583–596.

Cawse, J.N., Baerns, M., and Holena, M. (2004). Efficient discovery of nonlinear dependencies in a combinatorial catalyst data set. *J. Chem. Inf. Comput. Sci.* 44, 1, 143–146.

Clerc, F., Lenglitz, M., Farrusseng, D., Mirodatos, C., Pereira, S.R.M. and Raotomala, R. (2005). Library design using genetic algorithms for catalyst discovery and optimization, *Rev. Sci. Instrum.* 76, 6 pp. Article Number: 062208-1 to 13.

Corma, A., Serra, J.M., Sema, P. and Moliner, M. (2005). Integrating high-throughput characterization into combinatorial heterogeneous catalysis: unsupervised construction of quantitative structure/property relationship models. *J. Catal.* 232, 2, 335–341.

Corma, A., Serra, J.M., Serna, P., Valero, S., Argente, E. and Botti, V. (2005). Optimization of olefin epoxidation catalysts applying high-throughput techniques and genetic algorithms assisted by artificial neural networks. *J. Catal.* 229, 2, 513–524.

Cundari, T., Jun, D. and Yong, Z. (2001). Design of propane ammoxidation catalysts using artificial neural networks and genetic algorithms. *Ind. Eng. Chem. Res.* 40, 23, 5475–5480.

Dorigo, M. and Stützle, T. (2004). Ant Colony Optimization, MIT Press, Cambridge (MA), pp. 319.

Eckhard, K., Schlueter, O., Hagen, V., Wehner, B., Erichsen, Schuhmann, W. And Muhler, M. (2005). Spatially resolved mass spectrometry as a fast semi-quantitative tool for testing heterogeneous catalyst libraries under reducing stagnant-point flow conditions. *Appl. Catal. A: General* 281, 1–2, 115–120.

Farrusseng, D. and Clerc, F. (2007). Diversity management for efficient combinatorial optimization of materials, *Appl. Surf. Sci.* 254, 3, 772–776.

Feoktistov, V. (2005). Differential Evolution in Search of Solutions, Springer, Berlin, pp. 195.

Gobin, O.C. and Schueth, F. (2008). On the suitability of different representations of solid catalysts for combinatorial library design by genetic algorithms, *J. Comb. Chem.* 10, 6, 835–846.

Grubert, G., Kolf, S, Baerns, M., Vauthey, I., Farrusseng, D., van Veen, A.C., Mirodatos, C., Stobbe, E.R. and Cobden, D. (2006). Discovery of new catalytic materials for the water–gas shift reaction by high-throughput experimentation, *Appl. Catal. A: General* 306, 17–21.

Hastie, T., Tibshirani, R. and Friedman, J. (2001). The Elements of Statistical Learning, Springer, Berlin, pp. 552.

Huang, K., Zhan, X., Chen, F. and Lue, D.W. (2003). Catalyst design for oxidative methane coupling by using an artificial neural network and a hybrid genetic algorithm. Chemical Engineering Science 58, 1, 81–87.

Kennedy, J. and Eberhart. R.C. (2001). Swarm Intelligence, Morgan Kaufmann, San Francisco, pp. 512.

Larrañaga, P. and Lozano, J.A. (2002). Estimation of Distribution Algorithms, Kluwer, Boston, pp. 382.

McLachlan, G. And Peel, D. (2000). Finite mixture models. Wiley, pp. 419.

Maier, W.F., Stoewe, K. and Sieg, S. (2007). Combinatorial and high-throughput material science. *Angew. Chem. Int. Ed.* 46, 32, 6016–6067.

Moehmel, S., Steinfeldt, N., Engelschalt, S., Holena, M., Kolf, S., Baerns, M., Dingerdissen, U., Wolf, D., Weber, R. and Bewersdorf, M. (2008). New catalytic materials for the high-temperature synthesis of hydrocyanic acid from methane and ammonia by high-throughput approach, *Appl. Catal. A: General* 334, 1–2, 73–83.

Oh, K.S., Kim, D.K., Maier, W.F. and Woo, S.I. (2007). Discovery of new heterogeneous catalysts for the selective oxidation of propane to acrolein. *Comb. Chem. High Throughput Screening* 10, 1, 5–12.

Olong, N.E., Stoewe, K. and Maier, W.F. (2007). A combinatorial approach for the discovery of low-temperatue soot oxidation catalysts. *Appl. Catal. B: Environmental* 74, 1–2, 19–25.

Paul, J.S., Janssens, R., Debayer, J.M., Joeri, F.M., Baron, G.V. and Jacobs, P.A. (2005). Optimization of MoVSb oxide catalyst for partial oxidation of isobutane by combinatorial approaches, *Comb. Chem.* 7, 3, 407–413.

Rodemerck, U., Baerns, M., Holena, M. and Wolf, D. (2003). Application of a genetic algorithm and aneural network for the discovery and optimization of new solid catalytic materials. *Appl. Surf. Sci.* 223, 1–3, 168–174.

Saalfrank, J.W. and Maier, W.F. (2004). Directed evolution of noble-metal free catalysts for the oxidation of CO at room temperature, *Angew. Chem. Int. Ed.* 43, 15, 2028–2031.

Schölkopf, B. and Smola, A.J. (2002). Learning with Kernels, MIT Press, Cambridge (MA), pp. 644.

Serra, J.M., Baumes, L.A., Moliner, M., Serna, P. and Corma, A. (2007). Zeolite synthesis modelling with support vector machines: a combinatorial approach, Comb. Chem. High Throughput Screening, 10, 13–24.

Stanton, M. and Holcombe, J. (2007). Array-based split-pool combinatorial screening of potential catalysts. *J. Comb. Chem.* 9, 3, 359–365.

Strasser, P., Fan, Q., Devenney, M., Weinberg, W.H., Liu, P. and Norskov, J. (2003). High throughput experimental and theoretical predictive screening of materials – A comparative study of search strategies for new fuel cell anode catalysts. *J. Phys. Chem. B* 107, 40, 11013–1021.

Tompos, A., Hegedűs, M., Margitfalvi, J.L., Szabó, E.G. and Végvári, L. (2008). Multicomponent Au/MgO catalysts designed for selective oxidation of carbon monoxide: Appication of a combinatorial approach. *Appl. Catal. A: General* 334, 1–2, 348–356.

Tompos, A., Margitfalvi, J.L., Hegedu, M., Szegedi, A., Fierro, J.L.G. and Rojas, S. (2007). Characterization of trimetallic Pt-Pd-Au/CeO_2 catalysts combinatorially designed for methane total oxidation, *Comb. Chem. High Throughput Screening* 10, 1, 71–82.

Umegaki, T., Watanabe, Y., Nukui, N., Omata, K. and Yamada, M. (2003). Catalyst optimization for methanol synthesis by a combinatorial approach using a genetic algorithm assisted by a neural network. *Energy Fuels* 17, 4, 850–856.

Urschey, J., Kuehnle, A. and Maier, W.F. (2003). Combinatorial and conventional development of novel dehydrogenation catalysts. *Appl. Catal. A: General* 252, 1, 91–106.

Valero, S., Argente, E., Botti, V., Serra, J.M., Serna, P., Moliner, M., Corma A. (2009. DoE framework for catalyst development based on soft computing techniques, *Comput. Chem. Eng.* 33, 1, 225–238

Vapnik, V. (1998). Statistical Learning Theory, Wiley, Chichester, pp. 736.

Végáry, L., Tompos, A., Gobolos, S. and Margitfalvi, J. (2003). Holographic research strategy for catalyst library design: Description of a new powerful optimisation method, *Catal. Today* 81, 3, 517–527.

Watanabe, Y., Umegaki, T., Hashimoto, M., Omata, K. And Yamada, M. (2004). Optimization of Cu oxide catalysts for methanol synthesis by combinatorial tools using 96-wells microplates, artificial neural network and genetic algorithm. *Catal. Today* 89, 4, 455–464.

Yamada, Y. and Kobayashi, T. (2006). Utilization of combinatorial method and high throughput experimentation for development of heterogeneous catalysts. *J. Jpn. Pet. Inst.* 49, 4, 157–167.

Index

www.ingramcontent.com/pod-product-compliance
Lightning Source LLC
Chambersburg PA
CBHW050628190326
41458CB00008B/2183